GREATER Kansas City

PRESENTED TO:

PHOTO BY SCOTT INDERMAUR

GREATER Kansas City
UNLIMITED POSSIBILITIES

EDITOR	Rob Levin
PUBLISHER	Barry Levin
ASSOCIATE PUBLISHERS	John Lorenzo, Bob Sadoski
COMMUNITY LIAISON	Pam Whiting
C.O.O.	Renée Peyton
ASSOCIATE EDITOR	Rena Distasio
PROJECT DIRECTOR	Cheryl Sadler
PROJECT COORDINATOR	Muriel Diguette
PHOTO EDITORS	Ann Fowler, Jill Dible
WRITERS	Kimberly DeMeza, Rena Distasio, Grace Hawthorne, Regina Roths, Gail Snyder
COPY EDITOR	Bob Land
DESIGN MANAGER	Ann Fowler
BOOK DESIGN	Jill Dible and Compōz Design
JACKET DESIGN	Kevin Smith
PREPRESS	Vickie Berdanis
PHOTOGRAPHERS	Thomas S. England, Eric Francis, Scott Indermaur, Dennis Keim, Mario Morgado, Alan Weiner

Copyright © 2007 by Bookhouse Group, Inc.

Printed and bound in China

All rights reserved. No part of this book may be reproduced or transmitted in any form or by any means, electronic or mechanical, including photocopying of records, or by any information storage and retrieval system without permission in writing from Bookhouse Group, Inc.

RIVERBEND BOOKS
A division of BOOKHOUSE GROUP, INC.

Published by Riverbend Books
an Imprint of Bookhouse Group, Inc.
818 Marietta Street, NW
Atlanta, Georgia 30318
www.riverbendbooks.net
404.885.9515

Library of Congress Cataloging-in-Publication Data
Kansas City : unlimited possibilities / [editor, Rob Levin].
p. cm.
ISBN 978-1-883987-31-2
1. Kansas City (Kan.)—Description and travel. 2. Kansas City (Kan.)—Pictorial works.
3. Kansas City (Kan.)—Economic conditions. 4. Business enterprises—Kansas—Kansas City.
5. Kansas City Region (Kan.)—Description and travel.
I. Levin, Rob, 1955- F689.K2K34 2007
978.1'39034—dc22
2007011772

PHOTO BY THOMAS S. ENGLAND

The annual Jazz and Blues Festival has been held in or around Kansas City, Missouri, for nearly a quarter of a century, but for a few years took place in other venues. The most recent festival marked a return to the event's Kansas City roots, accommodating a crowd of thirty thousand fans and twenty-four acts over a two-day period. The bustling crowd is always hungry, mainly for barbecue. "Barbecue is not just a big thing here, it's the thing," said Betsy Donnelly, media director. "You have to be a carnivore in Kansas City." The festival is owned and operated by Mark Valentine, musician and event planner. "It was because of his personal passion, drive, and desire to see Kansas City reconnect with its musical roots that he brought the festival back. It's been really well received," said Donnelly.

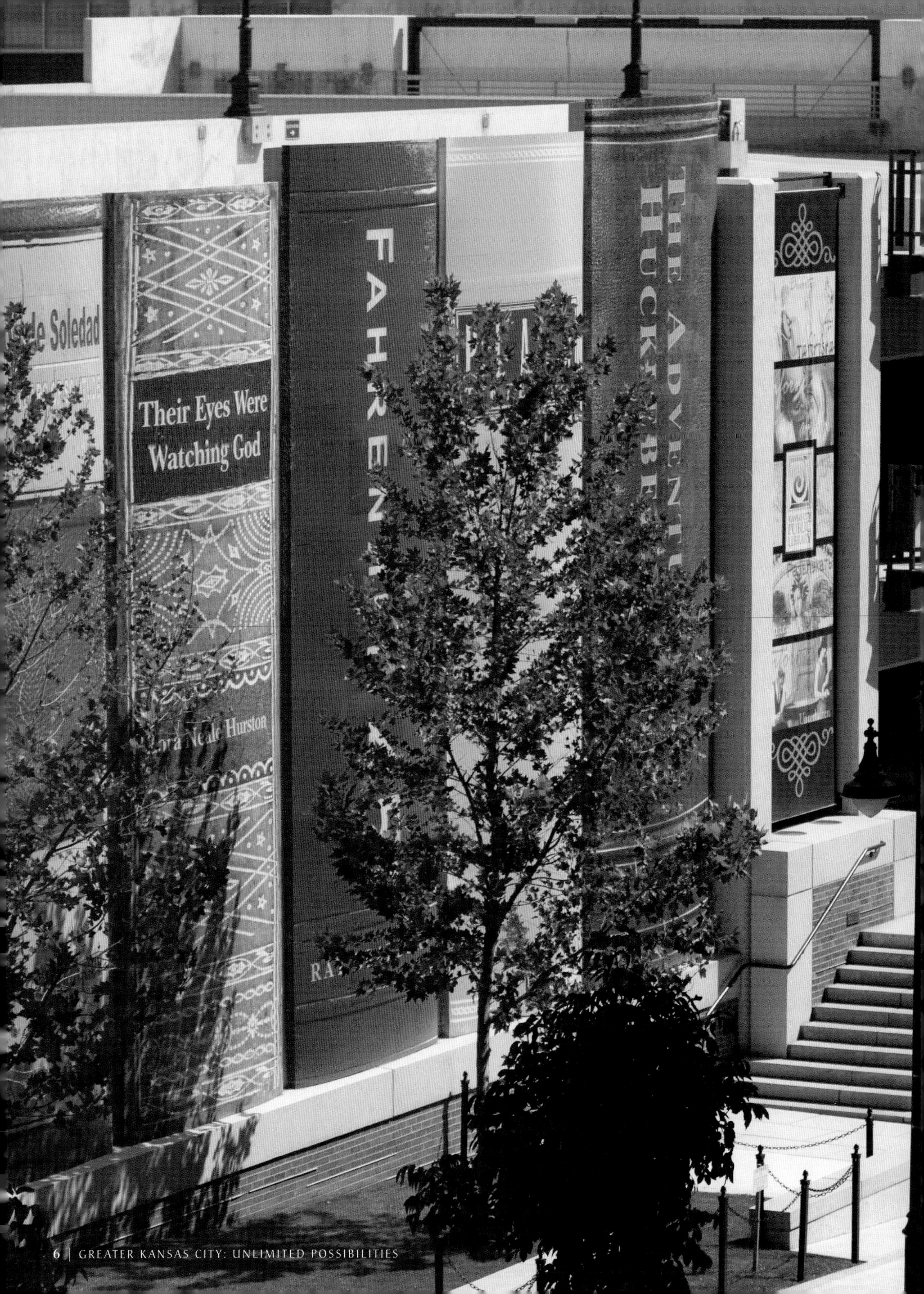
Their Eyes Were
Watching God

PHOTO BY THOMAS S. ENGLAND

When the Kansas City Public Library was first opened in 1897, it not only provided a resource for education, but also an "alternative to bawdy entertainment, gambling, and drinking," which according to an online history were readily available downtown. More than a hundred years later, in a move to stimulate downtown redevelopment, the Downtown Council, Greater Kansas City Community Foundation, and the Downtown Central Library purchased and renovated the old First National Bank Building. Subsequently JE Dunn Construction became involved in this and other downtown revitalization projects. In a bold move, citizens were asked to pick influential books that represented Kansas City. The titles became "bookbindings" for the exterior of the library's parking garage, which is, without a doubt, the most interesting parking garage you're likely to see.

Among the most recognizable pieces of modern sculpture in Kansas City are the giant shuttlecocks at the Kansas City Sculpture Park, which is a collaboration of the Nelson-Atkins Museum of Art, the Hall Family Foundation, and the city's Board of Parks and Recreation Commissioners. Each of the sculptures weighs fifty-five hundred pounds, stands nearly eighteen feet tall, and has a diameter of sixteen feet. They were designed by husband-and-wife team Claes Oldenburg and Coosje Van Bruggen.

Contents

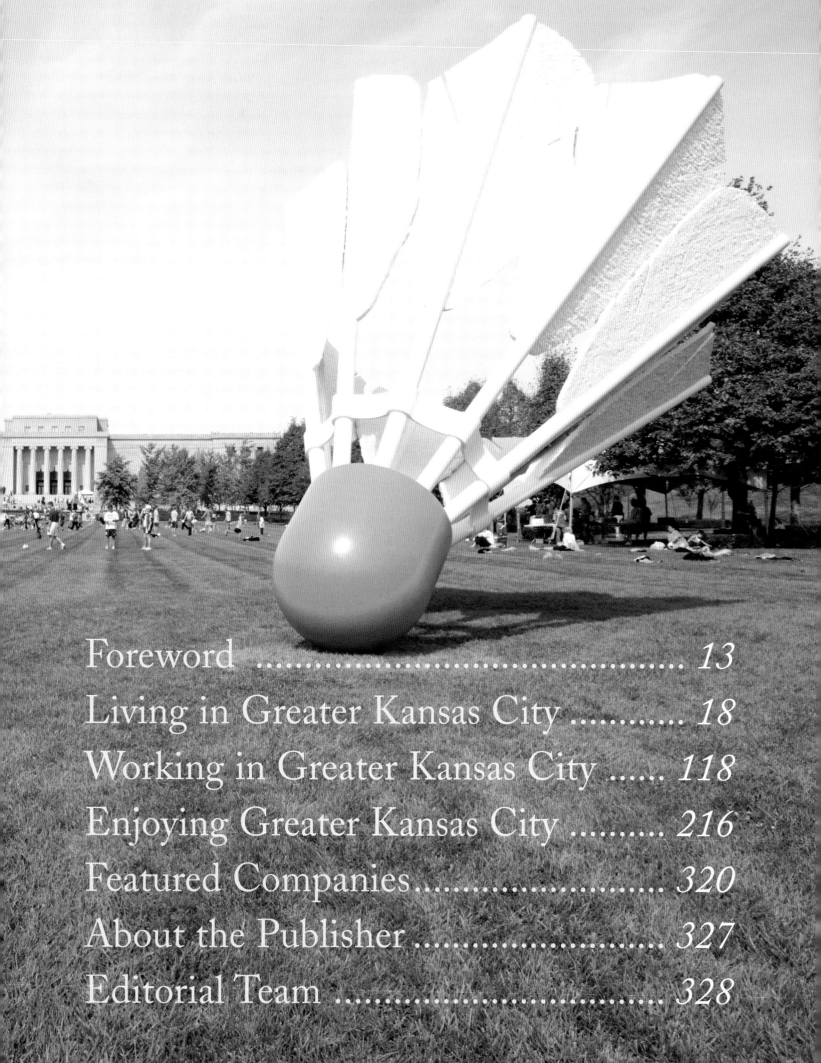

PHOTO BY SCOTT INDERMAUR

Foreword 13
Living in Greater Kansas City 18
Working in Greater Kansas City 118
Enjoying Greater Kansas City 216
Featured Companies 320
About the Publisher 327
Editorial Team 328

From fascinating, traditional favorites to invigorating, contemporary works, the Kansas City Ballet's performances have delighted audiences throughout the Midwest since 1957. Dancers trained in some of the world's most prestigious programs make up the Kansas City Ballet, bringing their dedication and talents to the stage—both at home and throughout the country.

The first roller-coaster riders who paid a nickel each to ride the Switchback Railroad at Coney Island in 1884 were thrilled to speed down a gradual fifty-foot drop at six miles an hour. In their wildest dreams they could not have imagined sitting atop the Spinning Dragons, a five-story-tall coaster with cars that spin as they move through hairpin turns and reach thirty miles per hour. This ride is just one of many at Worlds of Fun Amusement Park, which has been providing thrills for more than thirty-five years. So get your hands over your head, close your eyes, here we go. . . .

Foreword

GREATER KANSAS CITY: UNLIMITED POSSIBILITIES

The first written record of the "possibilities" that the Greater Kansas City area offers came from Captains Meriwether Lewis and William Clark. On June 26, 1804, the explorers camped at the confluence of the Missouri and Kansas rivers and noted that the site offered possibilities as a military outpost.

The settlers, lumber barons, railroaders, and others who followed Lewis and Clark saw the possibilities as well. Today, Greater Kansas City is a sprawling, vibrant community of nearly 2 million people—a community of rolling hills, sparkling fountains, and tree-lined boulevards, a community powered by an entrepreneurial spirit illustrated by names like Hall, Kauffman, Bloch, Stowers, and more.

A community whose possibilities remain . . . unlimited.

It's an exciting time to be in Kansas City. Possibilities are becoming reality: the Sprint Center, the Nelson-Atkins Art Museum, the Performing Arts Center, the Power & Light District, new projects and development from Olathe and Wyandotte County to the Northland and Lee's Summit.

The Greater Kansas City Chamber of Commerce is delighted to present this photographic journey through our two-state community. In these pages, you'll get a sense of the heart and soul of the place we call home.

The Chamber has worked on behalf of Kansas City business since 1887. We are committed to the growth and health of our two-state region, helping businesses large and small become smarter, more visible, and better connected, and working with elected officials and government at the local, state, and federal levels to meet the needs of our community.

This book celebrates all that Greater Kansas City has to offer: its history, its landscape, and its people.

Enjoy . . .

Peter S. Levi, president
Greater Kansas City Chamber of Commerce

would not have been possible without the support of the following sponsors:

AMERICAN CENTURY INVESTMENTS • BANK MIDWEST • BKD LLP • BLACK & VEATCH • BLACKWELL SANDERS PEPER MARTIN LLP • BLUE CROSS AND BLUE SHIELD OF KANSAS CITY • BURNS & MCDONNELL • CHASE SUITES HOTEL • CITY OF BLUE SPRINGS • COURTYARD BY MARRIOTT COUNTRY CLUB PLAZA • EWING MARION KAUFFMAN FOUNDATION • GARMIN INTERNATIONAL, INC. GOULD EVANS ASSOCIATES • GREATER KANSAS CITY CHAMBER OF COMMERCE • HOEFER WYSOCKI ARCHITECTS, LLC • HOTEL PHILLIPS • JE DUNN CONSTRUCTION • KANSAS CITY MARRIOTT DOWNTOWN • KANSAS CITY POWER & LIGHT • KANSAS CITY SYMPHONY • KANSAS CITY UNIVERSITY OF MEDICINE AND BIOSCIENCES • SPRINT • SWOPE COMMUNITY ENTERPRISES

14 | GREATER KANSAS CITY: UNLIMITED POSSIBILITIES

On this night, the historic Uptown Theater was the site for the Local Music Playoffs, an annual event hosted by the Web-based promotional site Bands Across Kansas City. Tomorrow, it might be a wedding, business meeting, or holiday party that brings in the crowds. Built in 1928 and located at 3700 Broadway (Broadway and Valentine) in midtown Kansas City, the Uptown hosts some of the world's best musical, theatrical, and comedic talent and is also a popular venue for all manner of special events. Austrian-born John Eberson originally designed the Uptown to replicate a romantic outdoor Mediterranean courtyard complete with classical arches and columns, seascapes, and a ceiling filled with twinkling stars, clouds, and mechanical flying birds. In 1994, the building underwent a $15 million renovation that restored its former colorful glory and added thirty-three thousand square feet of space to include four meeting and banquet rooms, kitchens, beverage service areas, expanded lobby space, and office suites.

Leading one of the National Football League's most spirited fan bases, the Kansas City Chiefs Cheerleaders know how to make some noise for their team. Working a packed Arrowhead Stadium, home of the Chiefs and one of the loudest stadiums in the league, the cheerleaders epitomize the spirit of Kansas City. During home games, when the stadium becomes a sea of red, noise levels can sometimes reach over 120 decibels, equal to that of an airliner taking off.

Living in Greater Kansas City

Strike up the band and join the parade. Since 1925, folks in Kansas City have looked forward each year to the American Royal Parade. With gorgeous floats, dozens of high school marching bands, giant balloons, saddle clubs, and, of course, clowns, there is plenty to see along the parade route down Grand Boulevard. If you are lucky, you might even witness the Pit Parade, hundreds of decorated, rolling barbecue pits. This premier fall event and related activities such as a rodeo and horse show has an annual economic impact on Kansas City of more than $60 million.

PHOTO BY SCOTT INDERMAUR

Life in Greater Kansas City is an exciting, eclectic experience, where neighborhoods are as diverse as the people who call them home. From the quickly rejuvenating downtown area, with its loft living and art galleries, to Hyde Park, where wagons once rallied and turn-of-the-century homes still stand, Kansas City living offers something for everyone. There's Brookside, with its laid-back lifestyle just south of Crown Center, and Historic Midtown, housing the fun and funky Westport. Head north toward the airport to find neighborhoods like Briarcliff, filled with some of the best of the old and new. Or step over the state line to Kansas, where it's all about trendy corporate suburbs.

In Kansas City, life pulses to the flow of cascading fountains and smooth jazz. With more than ninety eateries serving up barbecue rubbed and slow roasted, it's a place where competitions can be as hot as the sauce . . . where days can be spent at a park or zoo, monument or museum, or shopping to your heart's content . . . and where nights are filled with culture and entertainment served up by artists both homegrown and world renowned.

But no matter where you live, eat, or play, Kansas City offers a quality of life that is firmly rooted in its reputation as a regional stronghold for excellence in education and health care.

The region remains a place of educated, forward thinkers thanks to nearly two dozen colleges and professional schools, ranging from the Kansas City Art Institute to Park University to Johnson County Community College. As Kansas City strengthens its role in the life sciences realm, the Kansas City University of Medicine and Biosciences is continually evolving its courses to keep tomorrow's researchers and scientists at the cutting edge.

For over a century, the University of Kansas Hospital has been undertaking the challenge of educating tomorrow's health professionals. Serving as a training ground for scores of health professionals throughout the region and the world, this newly reorganized hospital is making substantial investments in technology to ensure continuing improvements in care for the area's residents. In addition, it is a keystone of a health-care community that includes historic staples like Bayer CropScience and Bayer Health Care, LLC, Children's Mercy Hospitals, Truman Medical Center, and Blue Cross and Blue Shield of Kansas City.

Distinctive and diverse, Kansas City offers a unique blend of lifestyles—from its vibrant core to its comfortable suburbs—that make it a uniquely attractive place to live.

AMERICAN DIGITAL SECURITY	50
AQUILA	64
BAYER CROPSCIENCE	47
BAYER HEALTHCARE, LLC	46
BLUE CROSS AND BLUE SHIELD OF KANSAS CITY	54
CHILDREN'S MERCY HOSPITALS	60
COLLIERS TURLEY MARTIN TUCKER	40
EWING MARION KAUFFMAN FOUNDATION	26
GREATER KANSAS CITY COMMUNITY FOUNDATION	74
JOHNSON COUNTY COMMUNITY COLLEGE	96
JOHNSON COUNTY GOVERNMENT	70
KANSAS CITY AREA DEVELOPMENT COUNCIL	30
KANSAS CITY POWER & LIGHT	42
KANSAS CITY REGIONAL ASSOCIATION OF REALTORS	92
KANSAS CITY UNIVERSITY OF MEDICINE AND BIOSCIENCES	34
KCMO WATER SERVICES DEPARTMENT	76
MARK ONE ELECTRIC COMPANY INC.	84
METROPOLITAN COMMUNITY COLLEGE	88
MIDWEST RESEARCH INSTITUTE	110
PARK UNIVERSITY	100
SWOPE COMMUNITY ENTERPRISES	104
TRUMAN MEDICAL CENTERS	82
THE UNIVERSITY OF KANSAS HOSPITAL	22

CHAPTER ONE: LIVING IN GREATER KANSAS CITY

It may have started small—one string of sixteen lights above a doorway in 1925—but there is nothing small now about the Annual Plaza Lighting Ceremony. More than eighty miles of lights and approximately 260,000 jewel-colored bulbs cover the expanse of the plaza for the opening of the holiday season on Thanksgiving evening. It is truly a magical time, whether you view the lights from a carriage or join the other three hundred thousand guests to applaud the lights and fireworks. But perhaps the most magical experience is to be the child selected at random from the audience to turn on the big switch on the stage.

The University of Kansas Hospital: A New Century, A New Level of Care

One step inside The University of Kansas Hospital and you'll find a place filled with a spirit of hope and healing, and teeming with patients, families, and health-care professionals.

As the region's premier academic medical center, The University of Kansas Hospital provides a full range of both inpatient and outpatient services, and through its affiliation with the University of Kansas Schools of Medicine, Nursing, and Allied Health, it participates in some of the nation's most exciting research and clinical trials.

But it was a different world back in 1998 when, after nearly a century of providing care, the hospital was in decline and rocked by a heart transplant debacle. Its reputation left the hospital unable to attract personnel, patients, or payors.

▲ *A commitment to patient-focused service has led to record growth and awards for quality care, including being the first and only hospital in the state of Kansas to achieve the coveted national Magnet designation for nursing excellence.*

CHAPTER ONE: LIVING IN GREATER KANSAS CITY

Determined to turn things around, new leadership led an initiative to separate the hospital operationally from the state system, leaving it without state dollars, no endowment, and only ten days of reserve operating funds.

Undaunted, the hospital made its top priorities patient satisfaction and high-quality patient care. Over the years, the hospital has participated in countless quality improvement initiatives, including the 100,000 Lives Campaign, an effort by the Institute for Healthcare Improvement to eliminate needless hospital deaths.

The campaign led to the formation of partnerships and programs such as rapid response teams that quickly answer code-blue situations. As a result of programs like these, the hospital saw a nearly 40 percent reduction of mortality rates for heart attacks and strokes, significantly lowered its incidences of infection and pneumonia, and completely transformed its bedside care.

Free to draw on both operational funds and bonds for capital investment, the hospital also has made many physical improvements.

In the fall of 2006, the hospital opened its Center for Advanced Heart Care, a $77 million, 238,000-square-foot facility that has removed any residue of the earlier heart scandal and brought the region some of the most advanced cardiac care and best patient outcomes found anywhere in the world.

The dedication to excellence shown in the heart center is also being applied to the hospital's new $37 million cancer center. Located in Westwood, this center encompasses 55,000 square feet and houses the area's largest outpatient cancer operation and most comprehensive breast cancer services.

As a result of its turnaround, the hospital has attracted a nursing staff that collectively holds the region's highest percentage of bachelor's degrees. Recently, the American Nurses Credentialing Center awarded the hospital Magnet designation in recognition of nursing excellence and outstanding patient care and outcomes. The hospital's initiatives have led it to an eleventh-place ranking in overall quality and safety among academic medical centers nationwide. Patient volume has risen to just under twenty thousand inpatient visits and nearly a quarter of a million outpatient visits, reflecting an eighty-thousand-patient increase over a four-year period.

It's truly a new dawn at The University of Kansas Hospital; it's the start of a new century of dedication to being a place where professionals want to practice and patients want to come for care. ♦

ONE STEP INSIDE THE UNIVERSITY OF KANSAS HOSPITAL AND YOU'LL FIND A PLACE FILLED WITH A SPIRIT OF HOPE AND HEALING.

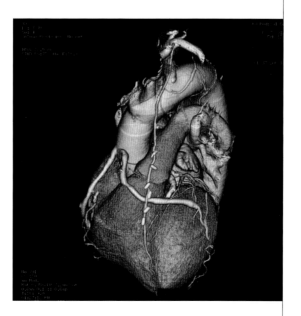

◀ *The University of Kansas Hospital's investment in the latest technology, visible in this dramatic heart image from its state-of-the-art sixty-four-slice scanner, is saving lives.*

24 | GREATER KANSAS CITY: UNLIMITED POSSIBILITIES

CHAPTER ONE: LIVING IN GREATER KANSAS CITY

PHOTO BY DENNIS KEIM

PHOTO BY DENNIS KEIM

Agriculture touches all our lives, from the food we eat to the clothes we wear, and the values, economy, and culture of our nation. The National Agricultural Center and Hall of Fame, or the Ag Center as it is known locally, was created by a federal charter to be the "national" museum of agriculture and a memorial to farming leaders. The main building houses the Hall of Fame, which recognizes leaders in science, education, government, and farm management—people like Thomas Jefferson, George Washington Carver, and John Deere. The Museum of Farming contains thirty thousand artifacts and a vast collection of antique farm machinery and implements. Visitors to Farm Town U.S.A. can explore on foot or ride the narrow-gauge train to visit the one-room schoolhouse, the blacksmith shop, general store, poultry hatchery, or the Morris railroad depot built in 1887. The center is entirely funded by private and corporate donations, and revenue generated on site.

The Kauffman Foundation: Advancing Entrepreneurship, Improving Education

Hailing from a Missouri farm and moving to Kansas City as a youth, Ewing Kauffman grew into an enterprising adult who possessed a great interest in changing the world. Kauffman's eye for opportunity led him to start a pharmaceuticals business in his basement that grew into a global enterprise with nearly $1 billion in annual sales and more than three thousand personnel.

While Kauffman's basement endeavor is a story of undeniable success, Kauffman's most enduring legacy to his community and the world is the Ewing Marion Kauffman Foundation. He established the foundation with the same sense of opportunity that he brought to his business endeavors and with the same convictions. Kauffman wanted his foundation to be innovative—to dig deep and get at the roots of issues to fundamentally change outcomes in people's lives. He wanted to help young people, especially those from disadvantaged backgrounds, achieve a quality education that would enable them to reach their full potential. He saw building enterprise as one of the most effective ways to realize individual promise

▲ *Minority entrepreneurs, like Ed Honesty (right), president and COO of Best Harvest Bakeries in Kansas City, Kansas, get business development coaching through the Kauffman Coaches program. This intensive coaching program identifies and works with minority entrepreneurs who have the resources and potential to significantly grow their businesses into large-scale enterprises.*

CHAPTER ONE: LIVING IN GREATER KANSAS CITY

and spur the economy. Today, the mission of the Kauffman Foundation mirrors Ewing Kauffman's commitment to fostering both ends of the opportunity continuum: education and entrepreneurship.

The foundation's work in K–12 education centers on mathematics, science, and technology. In 2005 the Kauffman Foundation began an ambitious ten-year initiative to better prepare students for careers of the twenty-first century through a focus on these fields. In addition to filling the workforce talent pipeline, students with strong math and science skills often become the world's most successful entrepreneurs. In two years, the foundation has committed more than $25 million for innovative programs and professional development offerings to help make Kansas City a national model for outstanding math and science education. Innovative and hands-on programs that were funded include FIRST Robotics, Project Lead the Way, and the JASON Project. These programs help students develop skills needed in industries and professions critical to the community's growth, such as engineering, medicine, and biotechnology.

In the entrepreneurial realm, the foundation works with educators, researchers, and public and private entities to develop and disseminate programs that support the next generation of entrepreneurs. The foundation is working with universities across the country in a number of ways. A growing number of universities have embraced the Kauffman Campuses initiative, which provides students in all fields of study the opportunity to access entrepreneurship courses. The foundation also works to advance innovation in America, providing education and developing an environment that enables university discoveries

(continued on page 28)

THE KAUFFMAN FOUNDATION BELIEVES IN LEADING WITH IDEAS TO MAKE DREAMS REAL.

▼ *The Ewing Marion Kauffman Foundation was established in the mid-1960s by the late entrepreneur and philanthropist Ewing Marion Kauffman. Based in Kansas City, Missouri, the Kauffman Foundation is the twenty-sixth-largest foundation in the United States, with an asset base of approximately $2 billion. The foundation focuses its grant making and operations on two areas: advancing entrepreneurship and improving the education of children and youth.*

EWING MARION KAUFFMAN FOUNDATION | 27

PHOTO BY ERIC FRANCIS

◀ James McGirt, an academic coach with the Kauffman Scholars initiative, teaching a session on measurement. Kauffman Scholars is a comprehensive, multiyear program designed to help promising, yet challenged, low-income urban students in Kansas City prepare for and complete a college education.

(continued from page 27)

to move more efficiently into the commercial market. And the foundation invests heavily in high-caliber research to study and better understand the dynamics of entrepreneurship and the importance it plays in the American and global economies.

Indeed, the Kauffman Foundation's entrepreneurship reach is international, radiating out from Kansas City as a center for leading thought on the topic. As an example, a new fellowship program created and funded by the United Kingdom's Chancellor of the Exchequer Gordon Brown will send top students from the U.K. to the foundation in 2007 for an intense entrepreneurship education and internship experience.

During his lifetime, Ewing Kauffman took the unconventional path to enrich others. He created jobs and recast the economy of his hometown. He found from his life as an entrepreneur inspiration to shape his philanthropy. His foundation's efforts to expand social welfare will be measured by the advances made to change the lives of youngsters, accelerate the course of business formation, and solidify America's economic future. From its home in Kansas City, Missouri, the Kauffman Foundation believes in leading with ideas to make dreams real and, in the end, to make a world where people's lives go beyond their dreams. ♦

▼ The Kauffman Foundation's Kauffman Campuses initiative makes entrepreneurship education available across university and college campuses, enabling any student, regardless of field of study, to access entrepreneurial training. Here, Ray Ricker teaches the Entrepreneurship in Music class at Eastman School of Music, University of Rochester, one of the eight original Kauffman Campuses.

CHAPTER ONE: LIVING IN GREATER KANSAS CITY

Located ten minutes northwest of downtown Kansas City, Parkville, Missouri, is small-town America at its best. A population of about five thousand friendly folks, a standout liberal arts university, and bucolic neighborhoods are just a few of the factors that make Parkville a great place to live. Another of its charms is an active, close-knit community, whose center for outdoor fun is English Landing Park. Situated along the banks of the Missouri River, the park provides walkers, joggers, and picnickers with groomed trails, shelters, and playground equipment. The park is also the location for a year-round calendar of festivals and outdoor events, including the highly popular Parkville Microfest beer festival and the Jazz/Blues and Fine Arts River Jam.

PHOTO BY DENNIS KEIM

Kansas City Area Development Council: ThinkKC as OneKC

Established in 1976, the Kansas City Area Development Council was one of the first regional economic development groups in the nation. The private, nonprofit organization has helped more than five hundred companies successfully locate new or expanded facilities to the eighteen-county, bi-state region. Together, these firms and organizations have created more than fifty thousand new jobs and utilized over 20 million square feet of office space.

Comprising a unique series of partnerships between more than two hundred corporate, city, county, and state leaders, KCADC's role is to actively promote the area to national and international businesses, position the region as a major competitor for top projects, equally represent all the communities, match the right company to the right community, and facilitate final negotiations between the company and its chosen location.

Critical to those efforts is a vision of Kansas City as a unified region, intertwined not only socially and culturally, but also economically. In 2004, KCADC launched a dual branding campaign designed to promote that vision. With an overall message of "we are one," the OneKC campaign is helping build unity and regional cooperation among area communities, counties, and corporations across the state line.

▲ *Every day, Kansas City competes aggressively with other major cities for jobs, capital investment, tax base, wealth creation, and the brightest workforce. As OneKC, it can compete and win. More than two hundred companies and community development organizations embrace the OneKC brand, serving as national ambassadors for the city's quality of life, professional workforce, and diverse economy.*

CHAPTER ONE: LIVING IN GREATER KANSAS CITY

The campaign presents Kansas City to corporate decision makers, site consultants, and real estate professionals as a seamless, unified "metropolitan product" of over 2 million people. By promoting this identity, KCADC increases the region's chances for new development and investments, resulting in more jobs and a higher tax base.

Launched concurrently with OneKC, ThinkKC is a national marketing and branding campaign designed to elevate Greater Kansas City as the preeminent location of choice for growing companies throughout the nation and the world—in other words, "ThinkKC" for major expansion, relocations, and investments.

Within only a few short years, both campaigns are helping promote the Kansas City region as a top-tier, world-class city. Since the campaign's kickoff, the area has undergone a nearly $7 billion building boom. Over two hundred corporations and communities in Kansas and Missouri have co-branded with KCADC and applied the red KC icon in their own marketing and advertising. In the first three years of the campaign, the council worked with its area partners in both states to fill over 4,821,308 total square feet of space and create 6,054 new jobs, and to generate $238 million in new payroll and $725 million in new capital investment for the region.

Promoting a region as unique as the Greater Kansas City area takes determination and vision. By developing public and private partnerships to promote the region as one city, the Kansas City Area Development Council not only encourages economic development, but also instills in its residents a renewed, unified sense of place and pride. ♦

THE KANSAS CITY AREA DEVELOPMENT COUNCIL INSTILLS IN ITS RESIDENTS A RENEWED, UNIFIED SENSE OF PLACE AND PRIDE.

PHOTO BY DENNIS KEIM

◀ In Kansas City, things get done the "OneKC Way," with regional cooperation and respect. It has allowed the city to attract more than six thousand new jobs to the community. OneKC is this generation's contribution to the growth and development of Kansas City. With fifty great cities, eighteen counties, and two states . . . together, KC does more.

CHAPTER ONE: LIVING IN GREATER KANSAS CITY

The American Royal is one of the largest combined livestock horse show rodeos in the nation. More than one hundred thousand spectators, like Tammy Stone and her nieces Francesca Fazzino and Victoria Stone, attend the annual parade. Sporting brand-new hats and trumpets, the girls are right in step with the spirit of the day. The American Royal organization provides scholarships, education, awards, and competitive learning experiences that reward hard work, leadership skills, and agrarian values—and it has been fulfilling that purpose since 1899. The organization reaches more than fifty-five thousand children annually with educational programs and events through the American Royal Museum, school tours, and other programs.

Kansas City University of Medicine and Biosciences: Preparing Tomorrow's Life Sciences Professionals

In the fast-evolving field of life sciences, Kansas City University of Medicine and Biosciences is preparing physicians, scientists, and researchers to be the instruments of tomorrow's change.

A private institution founded in 1916, KCUMB has an esteemed history of producing caring, compassionate physicians through its College of Osteopathic Medicine, a school recognized for its leadership in integrating the basic and clinical sciences at the onset of medical education through its Genesis Curriculum.

Today, as one of eight founding members of the Kansas City Area Life Sciences Institute, KCUMB has expanded to include a College of Biosciences, offering graduate-level education in biomedical sciences and bioethics.

The addition reflects KCUMB's commitment to meeting ever-changing industry and community needs, a mission more readily fulfilled by the university's unique mutability. "As a smaller, private institution, we have the ability to respond more quickly to changes in the health-care environment

▲ *Norbert Seidler, Ph.D., professor and chair of biochemistry, performs an experiment in his KCUMB laboratory. His discovery that the formation of thin, plasticlike fragments in the brain may contribute to diseases such as Alzheimer's has prompted excitement in the research community.*

CHAPTER ONE: LIVING IN GREATER KANSAS CITY

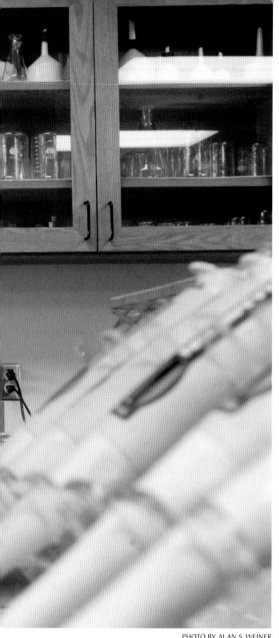

PHOTO BY ALAN S. WEINER

and to needs that exist in our community," explains president and chief executive officer Karen L. Pletz, J.D.

KCUMB's responsiveness is also evident in programs like its master's of arts in bioethics, which provides students a greater understanding of how to deal with complex ethical dilemmas. "There are so many moral enigmas today," says Pletz. "Bioethics endeavors to provide a framework for those very difficult medical and moral dilemmas that face physicians and researchers in today's environment."

In recognition of the strong connection between beliefs and healing, KCUMB has also implemented a comprehensive Spirituality in Medicine program, twice recognized with the John Templeton Spirituality and Medicine Curricular Award. "Research shows that a patient's spiritual beliefs impact the ability to heal," explains Pletz. "So we felt it was critical to teach our students an enhanced awareness of this impact and how to apply that greater understanding in medical care."

Seeing the need for physicians to guide the future of health care, KCUMB and Rockhurst University partnered to develop a unique master's of business administration in health care leadership. "People who are most knowledgeable about caregiving and about research should be prepared to take an active and well-educated role in impacting health policy," says Pletz. "This program gives our graduates that edge and will have a huge positive impact on medicine in years to come."

The university's impact will also be felt in activities taking place through the new Dybedal Center for Research, a forty-five-thousand-square-foot facility housing laboratories for Biosafety levels I and II research, as well as the Dybedal Clinical Research

(continued on page 36)

"WE CONSIDER IT A VERY IMPORTANT PART OF OUR MISSION TO SERVE THE COMMUNITY."

▼ *Diane Karius, Ph.D., director of the Kesselheim Center for Clinical Competence, monitors students' responses to scenarios from a control room at the Kesselheim Center for Clinical Competence. Students gain valuable experience in clinical situations through working with high-tech human patient simulators.*

PHOTO BY ALAN S. WEINER

KANSAS CITY UNIVERSITY OF MEDICINE AND BIOSCIENCES | 35

KANSAS CITY UNIVERSITY OF MEDICINE AND BIOSCIENCES

▲ Michael Dempewolf, a second-year KCUMB medical student, takes a child's blood pressure reading during a Score 1 for Health screening at Scuola Vita Nuova in Kansas City, Missouri. Score 1 for Health, sponsored by KCUMB and the Deron Cherry Foundation, annually provides health screenings to more than sixteen thousand elementary-age children living in the urban core.

◀ Medical students take a relaxing break from their studies at one of several picnic tables on the picturesque KCUMB campus.

(continued from page 35)

Center, the only adult academic clinical research center in Kansas City, Missouri.

Additionally, KCUMB is home to the Kesselheim Center for Clinical Competence, the area's first comprehensive human patient simulation center. Housing eight patient simulators, ranging from infants to adults, the center enhances medical students' early clinical training by giving them the opportunity to experience more than eighty complex combinations of emergent situations.

Together, the university's programs and facilities augment the teachings of a faculty whose lives are devoted to excellence. "Our faculty is dedicated not only to teaching the science of medicine and biomedical sciences, but also to teaching future physicians and researchers how to practice with the values and compassion that will advance patient care," says Pletz.

For KCUMB, dedication to the betterment of life also extends to the community. In physical presence, the university has been a catalyst for urban-core improvements, replacing unsightly properties with new structures or greenspace and helping to attract new business to the area. "We've strived to make the northeast urban core a much more vital, attractive, and meaningful place to be," says Pletz.

With programs like Score 1 for Health, cofounded by former Kansas City Chiefs football player Deron Cherry, KCUMB provides more than sixteen thousand children each year with free health screenings, while giving second-year medical students a chance to experience live patient interaction.

Through this type of hands-on involvement, faculty, staff, and students annually lend more than one hundred thousand hours to help causes in the Greater Kansas City region.

These activities, Pletz notes, are meant to make a lasting impact beyond graduation. "We consider it a very important part of our mission to serve the community," she says. "It's certainly integral to the education of our students, and we hope that it will foster a lifelong commitment in their personal lives as well." ♦

▲ *KCUMB medical student Valaree Smith, center, shares a morning cup of coffee with Benedictine sisters Lillian Harrington, left, and Marcia Ziska at Mount St. Scholastica in Atchison, Kansas. Smith and other KCUMB medical students participated in Igniting the Spirit, a unique course that prepares future physicians to appreciate the role an individual's spirituality plays in care giving and healing, especially in end-of-life care.*

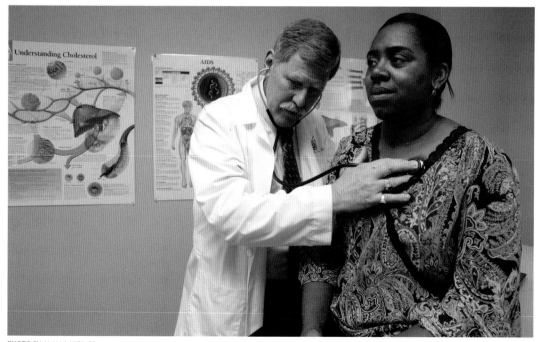

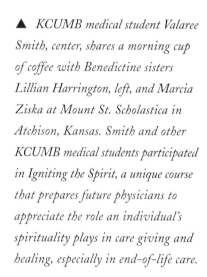

◀ *Richard Ogden, D.O., assistant professor of family medicine, examines a study participant at the Dybedal Clinical Research Center. Located on the KCUMB campus, the Dybedal Clinical Research Center is the first adult, not-for-profit, academic clinical research center in Kansas City, Missouri, and has become an emerging leader in advancing medical knowledge through clinical research.*

PHOTO BY ERIC FRANCIS

PHOTO BY ERIC FRANCIS

CHAPTER ONE: LIVING IN GREATER KANSAS CITY

At the Legends at Village West, shopping and entertainment combine with education to create a unique experience. Woven among the plethora of retail, dining, and entertainment options is a self-guided, audio walking tour of more than eighty Kansans noted for their contributions. From athletics and art to science and politics, the village pays tribute to such notables as golfing great Tom Watson, silent film star Buster Keaton, airplane pioneer Clyde Cessna, and senatorial leader Bob Dole. Housed in the 750,000-square-foot village is an array of lush and native landscaping features, architectural elements, statues and medallions, and courtyards with features like a 150-foot brick smokestack, a 60-foot tiered fountain, and more. A host of restaurants and eateries includes the nation's first T-Rex, a new venture by the creator of the popular Rainforest Café.

PHOTO BY ERIC FRANCIS

GREATER KANSAS CITY: UNLIMITED POSSIBILITIES | 39

Colliers Turley Martin Tucker Defines Market Leaders Committed to Client Service and Community

Who knows **The State of Real Estate®** in Kansas City? Colliers Turley Martin Tucker does. A leader in commercial real estate services, Colliers Turley Martin Tucker (CTMT) knows its business and its markets because of its unique approach founded on four key cornerstones: client solutions, committed employees, civic involvement, and constant growth.

Through a series of strategic mergers including the union of Kansas City–based Kerr Company and Zimmer-Steinbach in 1995, CTMT has grown to be the dominant commercial real estate firm in the Central United States, with regional offices in Kansas City, St. Louis, Cincinnati, Columbus, Dayton, Indianapolis, Minneapolis/St. Paul, and Nashville. In this Kansas City market, CTMT handles more transactions than any other third-party real estate firm in the city.

The employee-owned company offers a wide range of products and services as varied as the clients' needs, objectives, and goals—all of which are designed to create value. "We approach our clients' long-term needs as their partner. We listen carefully, and match our best talent from multiple disciplines to deliver the services," says S. Frazier Bell, managing principal. "And, because we support our personnel and our clients by continuous investment in research, information systems, technology, and training, we're confident that we're maximizing the success of our clients, our associates, our communities, and our firm." ♦

▲ *Colliers Turley Martin Tucker moved to 2600 Grand in Crown Center in 2006 to accommodate its growing staff and business needs. CTMT selected Crown Center as its new home because of its exceptional amenities, location, and rich Kansas City history.*

"WE'RE CONFIDENT THAT WE'RE MAXIMIZING THE SUCCESS OF OUR CLIENTS, OUR ASSOCIATES, OUR COMMUNITIES, AND OUR FIRM."

CHAPTER ONE: LIVING IN GREATER KANSAS CITY

The Johnson County Park and Recreation District has more than eighty thousand rounds of golf played annually on its two public facilities: Tomahawk Hills Golf Course, Shawnee, and Heritage Park Golf Course, Olathe.

Kansas City Power & Light: The Future of Energy, Today

Kansas City Power & Light has been lighting the way for Kansas City's growth and prosperity since its founding in 1882. From its humble beginnings as Kawsmouth Electric Company, KCP&L has become one of the leading regulated providers of energy products and services for homes, businesses, industries, and municipalities in the Midwest.

As the Kansas City area continues to grow, so does KCP&L. More than a century ago, it produced enough power for forty arc lamps from three miles of current-carrying lines. Today, more than two thousand KCP&L employees serve approximately half a million customers in twenty-four counties of western Missouri and eastern Kansas, a territory that includes approximately forty-six hundred square miles.

The company has a balanced generation mix that utilizes coal, nuclear, wind, and gas to provide power to customers and sell to the wholesale market, and has more than forty-one hundred megawatts of efficient generating assets in operation.

A leader in fuel procurement, as well as efficient power production and distribution, KCP&L has established a long-standing reputation for providing reliable, affordable energy to its customers. In fact, the utility

▲ *In 2006, KCP&L completed its 100.5-megawatt Spearville Wind Energy Facility. The sixty-seven wind turbines provide enough renewable electricity to serve the annual energy needs of approximately thirty-three thousand homes.*

company is one of the Midwest's most affordable energy suppliers. In national comparisons, it ranks among those utilities offering the lowest average prices for all customer segments and the least number of power interruptions due to outages.

**Supply and Demand:
The Importance of Balance**

Looking to the future, KCP&L developed a Comprehensive Energy Plan to meet the region's energy needs. "The comprehensive plan provides a clean, low-cost supply of electricity while protecting the customer from the high cost and volatility of natural gas–based generation," says Mike Chesser, chairman and CEO of Great Plains Energy, the parent company of KCP&L. "The plan will stimulate economic development, improve air quality, and incorporate renewable wind energy. It also provides customers with tools to manage their energy costs." During the planning stage, KCP&L reached out for input to hundreds of people in its service area, including employees, customers, industry experts, environmentalists, community leaders, and government officials. The plan features a portfolio of initiatives, such as environmental upgrades at existing power plants, alternative sources of energy, and applications for energy efficiency and demand-response programs.

Chesser says an important part of the company's energy plan involves its customers. KCP&L partners with residential and business customers to help them manage their energy usage and costs. With the Energy Optimizer program, KCP&L places the power to save at customers' fingertips. The program gives them the flexibility to adjust their home or business thermostat remotely, for convenience and cost-effectiveness. The Energy Analyzer is another interactive tool that helps customers

(continued on page 44)

KCP&L HAS ESTABLISHED A LONG-STANDING REPUTATION FOR PROVIDING RELIABLE, AFFORDABLE ENERGY TO ITS CUSTOMERS.

◀ *At KCP&L, we have a strong commitment to supplying the region with reliable electricity that customers have come to expect. KCP&L is ranked third in the nation for system reliability and first in the Midwest for storm response.*

KANSAS CITY POWER & LIGHT

◀ Since its founding in 1882, Kansas City Power & Light has been lighting the way for Kansas City's growth and prosperity.

(continued from page 43)

learn how they are using energy at home or their business, and offers ways to conserve energy and save money.

Wind Energy on the Horizon

KCP&L is taking steps to improve the environment by developing cleaner, greener energy sources. An energy facility in Spearville, Kansas, is harnessing one of nature's most plentiful resources: the wind. The Spearville Wind Energy Facility, completed in fall 2006, is the first large-scale wind facility in Kansas to be owned and operated by a regulated electric utility. The sixty-seven wind turbines, located on a five-thousand-acre facility, are capable of producing 100.5 megawatts of electricity, which is enough renewable electricity to serve the annual energy needs of approximately thirty-three thousand homes. ♦

◀ KCP&L provides more than the energy that powers our community. We empower our employees to make a difference, because we know that the mark of a great community is the upward mobility of all of its citizens.

CHAPTER ONE: LIVING IN GREATER KANSAS CITY

La luna glows like "a big pizza pie" from behind the spire of Figlio Tower, a relatively new addition to the historic Country Club Plaza in Kansas City, Missouri. A vision made reality by Jesse Clyde (J.C.) Nichols, the Plaza, the first suburban, open-air shopping center in the nation, is classified as "The Jewel of Kansas City." Nichols, a world traveler, admired the arts and architecture of Europe and filled the Plaza with Old World fountains, mosaics, murals, and sculpture, including an original piece by Italian sculptor Donatello Gabriella. Still a central shopping district of exclusive shops, international retailers, and entertainment, the Plaza's primarily Spanish architecture includes replicas of Spain's famous Giralda Tower and the Seville Light. The domed Figlio Tower with copper accents was built in the 1980s to draw attention to the northeast corner of the Plaza and adds ambience for the diners in the fine Italian restaurant that operates beneath it.

Bayer Animal Health: Science for a Better Life

Home to over one hundred companies dedicated to advancing the science of animal health and nutrition, Greater Kansas City has become known as the nation's Animal Health Corridor. With roots in the corridor since 1919, Bayer HealthCare's Animal Health Division leads the way in research and development of new products for companion animals and livestock.

Bayer Animal Health understands the significance of animals in our lives. In fact, many of us now demand the same quality of care for our four-legged friends as we do for ourselves. With a mission to protect animals and benefit people, Bayer Animal Health has become an industry leader in cost-effective, innovative solutions for preventing and treating disease. Its arsenal of products represents the latest advancements in flea, tick, and mosquito control; antimicrobials; and antibiotics.

Not only is Bayer Animal Health a member of the Animal Health Corridor, it is also a charter member and cofounder of the economic development initiative of the same name. The Animal Health Corridor initiative was established to not only help support existing animal health companies located in the Greater Kansas City area, but also to attract new ones to the region as well. Involvement in that effort is just one more way in which Bayer Animal Health continues to be a solid contributor to the future of animal health in Greater Kansas City and beyond. ♦

▲ *Access to the great outdoors is essential for a happy, healthy dog. Thanks to Bayer Animal Health's arsenal of products, dogs can romp and play without fear of compromising their health. The company's safe, effective anti–flea and tick products not only keep the pooches healthy, but their owners as well.*

BAYER ANIMAL HEALTH UNDERSTANDS THE SIGNIFICANCE OF ANIMALS IN OUR LIVES.

Bayer CropScience: Building Partnerships for Growth

To help achieve abundant and wholesome harvests in an affordable, sustainable, and environmentally sound way, farmers around the globe rely on the innovative technologies of Bayer CropScience.

The company's state-of-the-art manufacturing facility in Kansas City, Missouri, provides many of those technologies, producing some of the industry's most advanced herbicides, insecticides, and fungicides—products that help farmers fight damaging pests, plant diseases, and weeds that significantly decrease harvests.

A research-based crop science company, Bayer CropScience has its U.S. headquarters in Research Triangle Park, North Carolina. The U.S. company is a part of the global Bayer CropScience group, which is based in Monheim, Germany.

A leading partner in the global production of quality food, feed, and fiber to meet the needs of a growing world population, the company is committed to building its growth through innovation. Its R&D facilities across the globe enable the company to explore innovation opportunities with the goal of offering the right solutions to meet customers' specific needs. The company's research facility in Stilwell, Kansas, a part of the Greater Kansas City area, is an important part of that effort.

In Kansas City, Bayer CropScience continues to look to the future and is working to strengthen even further its production site's position as a world-class competitive manufacturing operation, while continuing to meet its obligations as a responsible and involved corporate citizen in the community. Its commitments in support of neighborhood, educational, and environmental initiatives remain important priorities for the company. ♦

▲ *The Bayer CropScience production operations in Kansas City are based on cutting-edge modern chemistry, the best and safest in manufacturing technology, and a highly trained world-class workforce. Here, A-Operator Gary Conard makes adjustments in one of the site's crop protection manufacturing processes.*

FARMERS AROUND THE GLOBE RELY ON THE INNOVATIVE TECHNOLOGIES OF BAYER CROPSCIENCE.

PHOTO BY SCOTT INDERMAUR

Whether it's a loft in the city center or a home in a laid-back suburb, Greater Kansas City's neighborhoods can make anyone feel right at home. From the River Market to the Crossroads Arts District, Kansas City's downtown is a place where something is always going on. The area is home to a number of boutiques and galleries, nightclubs and cafés, parks and museums, many within walking distance. For shopping, don't miss Crown Center, and for a night on the town there's the 18th & Vine Historic District. Southwest of the downtown area, there are places like Brookside, one of the hippest neighborhoods in town. Home of the chic and unique, Brookside is the place where houses and offices have preserved their architectural elegance. Brookside is located just up the street from Country Club Plaza, one of Kansas City's finest shopping addresses. Up north, toward the airport, the city has seen considerable residential growth, complementing years of commercial development that gives this area of town a full range of amenities for daily living. From single-family homes to apartments to mixed-use developments, dramatic change is taking place. If suburban living is more your style, head southeast to Blue Springs, a place with friendly streets, green parks, and growing commercial opportunities. Or step across the state line to Kansas, where life is a blend of corporations, tree-lined streets, and trendy retail.

48 | GREATER KANSAS CITY: UNLIMITED POSSIBILITIES

CHAPTER ONE: LIVING IN GREATER KANSAS CITY

This home on North Broadway, built in 1869 by Leavenworth entrepreneur E. B. Allen, began as a smaller, Italianate-style structure, but over time became more Greek Revival in style due to numerous additions.

PHOTO BY ALAN S. WEINER

American Digital Security: We're There When You're Not

While American Digital Security (ADS) is one of Kansas City's fastest-growing small companies, its long list of big business clients agree this company is a major player in the security industry. In business since 2002, ADS has grown from revenues of $750,000 in its first year of operation to over $3 million a year in 2006.

"Our formula for success has not changed much since we started and continues to be our driving force today," says Buddy Mason, owner. "By remaining true to our core values of working hard and providing personalized, quality service, we have watched our company become one that can quickly move on customers' requests, delivering customizable solutions that fit their budgets while still providing the security they need to be safe." ADS specializes in the design, system integration, installation, and service of closed-circuit television (CCTV), digital/network video surveillance, card access control, IP products, door intercom systems, and intrusion detection systems. The company also manufactures and distributes high-quality security products across North America.

Every ADS security system begins with a free on-site consultation by experienced security specialists. Trained field technicians complete installation and tailor every system to meet each client's specific needs.

▲ *Digital closed-circuit television allows for crystal-clear playback of footage and audio, helping to discern facial features, clothing and hair color, and voice characteristics. Digital also allows activities to be viewed simply by selecting date and time.*

CHAPTER ONE: LIVING IN GREATER KANSAS CITY

Once installed, the company provides assistance twenty-four hours a day, seven days a week, with a highly trained technical team ready to answer any questions and place clients firmly at ease.

With the resources and experience to equip and service any size business or agency, ADS counts among its clientele over four hundred grocery stores and food distributors, such as Price Chopper and Roberts Dairy, as well as several manufacturing plants, including Boulevard Brewing Company. Hospitals and police departments throughout the Midwest also benefit from ADS systems.

From a simple setup for the neighborhood convenience store to elaborate systems like that at the Golden Eagle Casino, where twenty digital video recorders capture images from over 320 cameras watching gaming tables, money cages, slot machines, and parking lots, ADS offers a versatile range of solutions. Regardless of system size, ADS clients are realizing a return on their investment through fewer losses, denial of fraudulent claims, increased employee productivity, and safer workplaces.

As the exclusive security provider for organizations like the Missouri School Boards' Association (MSBA), Associated Wholesale Grocers (AWG), and SYSCO foods, ADS is able to offer special programs and discounted pricing to members of these associations.

And because ADS places great importance on safety and security in educational institutions, it provides affordable systems through its school initiative program. Among the more than fifty facilities equipped to date are Missouri's Odessa R-VII School District and the University of Missouri–Kansas City (UMKC).

With technology always evolving, ADS consistently stays abreast of the latest innovations. No more grainy black-and-white videotapes; clients can review varied details such as physical features and license plate numbers and store the information for several months on end. Clients can also remotely view real-time information on their systems via their cellular phones, personal data assistants, or laptop computers. ♦

"BY REMAINING TRUE TO OUR CORE VALUES OF WORKING HARD AND PROVIDING PERSONALIZED, QUALITY SERVICE, WE CAN QUICKLY PROVIDE THE SECURITY OUR CUSTOMERS NEED TO BE SAFE."

PHOTO BY SCOTT INDERMAUR

◀ *Each security camera is mounted for optimum viewing of areas where theft may occur. To determine prime locations, American Digital Security begins with an on-site consultation prior to any installation.*

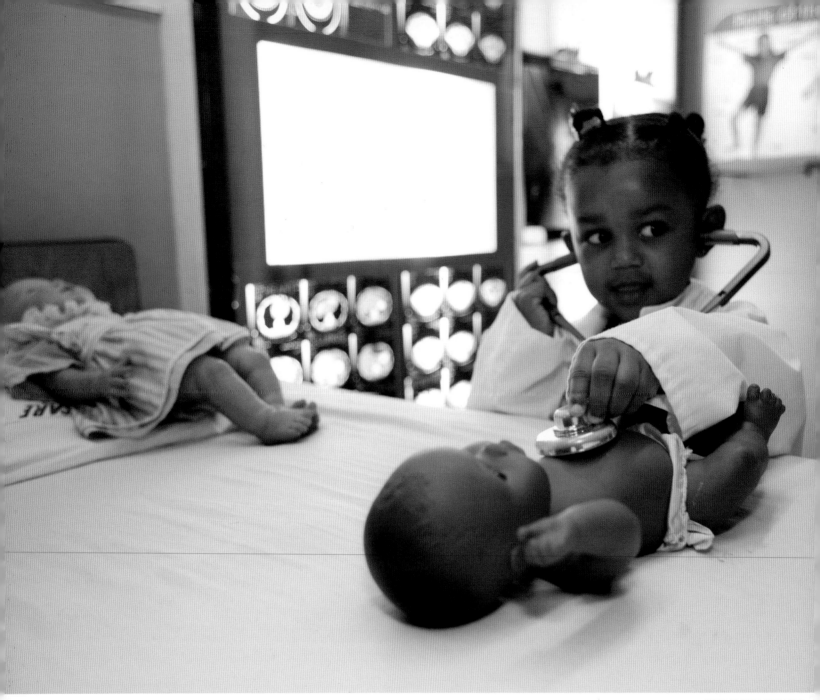

PHOTO BY SCOTT INDERMAUR

Wouldn't it be great if your job was to play all day? The Children's Museum of Kansas City provides an innovative, hands-on learning environment where young children can explore, discover, and use their imaginations. Sofia Torres and her friend Anna Borunda (right) are making the most of exhibits such as the Grocery Store, Slice of the City, Health and Safety, and Creation Station, while Olivia Coe (above) is busy examining a patient. From its original site in the historic carriage house of the Kansas State School for the Blind to its much-expanded present location, the Children's Museum is a favorite destination for residents and visitors alike.

CHAPTER ONE: LIVING IN GREATER KANSAS CITY

PHOTO BY SCOTT INDERMAUR

PHOTO BY SCOTT INDERMAUR

GREATER KANSAS CITY: UNLIMITED POSSIBILITIES | 53

Blue Cross and Blue Shield of Kansas City: Affordable, Dependable, Accessible

Blue Cross and Blue Shield of Kansas City, the community's only local, not-for-profit health insurer, has proudly served the Kansas City community since 1938. It is part of the national Blue Cross and Blue Shield system covering 98.5 million Americans.

Serving nearly nine hundred thousand members, Blue Cross and Blue Shield of Kansas City just experienced its sixth consecutive year of double-digit revenue and membership growth. Providing dependable, exemplary service and broad access to health-care providers is the primary key to its success in the market. To illustrate its strength in the market, the company adds three hundred new members to its health plans each day.

As the area's largest health insurer, Blue Cross and Blue Shield of Kansas City is unique among its competitors for offering health plans to every consumer category. It provides affordable plans for employer groups of every size, as well as individuals and families. Beyond traditional coverage, plan choices include dental, prescription, and Medicare supplements. For the smaller employer, the ChamberCHOICE plan offers a first-year renewal rate cap, which enables these employers to better budget their expenses over a two-year period.

▲ *Dr. Rafaela Herrera works with Blue Cross and Blue Shield of Kansas City in the Healthy Steps program. Healthy Steps is a national initiative focusing on the importance of providing appropriate care to children during the crucial first three years of life.*

CHAPTER ONE: LIVING IN GREATER KANSAS CITY

While its products and services cover members of the community in times of need, this company's core mission is to make a difference by improving the health of the communities it serves.

That's why Blue Cross and Blue Shield of Kansas City's business activities go beyond business to lead the way in providing its members with effective options for better health.

For instance, through Blue Cross and Blue Shield of Kansas City's unique, comprehensive health and wellness program called "A Healthier You," local employers are becoming more engaged in opportunities to help improve their employees' health.

Created in 2005 as a way to address an alarming trend of declining health found in the Blue Cross and Blue Shield of Kansas City market area and across the nation, "A Healthier You" is a high-touch, high-tech program, offering both personalized onsite programs and innovative online resources.

Through the program, Blue Cross and Blue Shield of Kansas City offers an on-site health fair and online Health Risk Appraisal to employer groups to help educate their employees on their health status and potential risks. Following these activities, on-site classes—such as smoking cessation, nutrition and weight management, and stress management—are offered, along with one-on-one counseling with a personal health coach, if needed. In addition, "A Healthier You" participants have access to a wealth of healthier lifestyle information, tools, and services through a secured Web site containing customized features such as personalized reminders, educational materials, and more.

Blue Cross and Blue Shield of Kansas City also makes a significant impact on its communities through charitable contributions, in-kind donations, and volunteer activities. Over three hundred community organizations are the beneficiaries of these

(continued on page 56)

> THIS COMPANY'S CORE MISSION IS TO MAKE A DIFFERENCE BY IMPROVING THE HEALTH OF THE COMMUNITIES IT SERVES.

◀ *Blue Cross and Blue Shield of Kansas City maintains a state-of-the-art fitness center for employees and their family members as part of its commitment to employee health. Employees have access to equipment, classes, and a personal trainer to guide them on their path to better health.*

PHOTO BY DENNIS KEIM

BLUE CROSS AND BLUE SHIELD OF KANSAS CITY

PHOTO BY MARK McDONALD

◀ Blue Cross and Blue Shield of Kansas City supports Kansas City Young Audiences, a group devoted to developing and inspiring children's creative talent and appreciation for the arts. Annually, this organization provides over eighteen hundred workshops in music, drama, dance, and visual arts to more than 165,000 young people in the Kansas City community.

(continued from page 55)

contributions each year. From member school districts to cultural organizations to commercial endeavors, Blue Cross and Blue Shield of Kansas City supports the entities that make a positive difference in its service areas. The company's community efforts were recognized in 2007 as it was honored as Business Philanthropist of the Year by the Kansas City Council on Philanthropy.

The company also works to strengthen the community through activities that promote healthy lifestyles, preventive health care, and better access to services.

For example, in recognition of the growing concern over childhood obesity, Blue Cross and Blue Shield of Kansas City has created a Community Wellness Initiative. Designed to help children and their families understand the importance of good nutrition and daily exercise during childhood and into adulthood, this program calls attention to children's and adults' health issues by assessing areas of need and

PHOTO BY MARIO MORGADO

◀ As part of its commitment to the Kansas City community, Blue Cross and Blue Shield of Kansas City and its employees participate in a variety of activities to benefit nonprofit organizations in the area. Here, Carla Jenkins, an employee in the IT division, braves the cold as part of the company's team running at the 2006 Jingle Bell Run to raise money and awareness for the Arthritis Foundation.

CHAPTER ONE: LIVING IN GREATER KANSAS CITY

◀ *A Healthier You is Blue Cross and Blue Shield of Kansas City's unique health and wellness program for area employers. A hands-on health fair and Health Risk Appraisal kicks off the comprehensive program for participating employers. Customized onsite classes and online tools and resources help employees lead healthier lives. Here, Tracy Foreman, a Blue Cross employee, guides an employee taking her Health Risk Appraisal.*

PHOTO BY SCOTT INDERMAUR

implementing creative solutions. In addition, Blue Cross and Blue Shield of Kansas City sponsors the Wellness Van, which makes weekly appearances around the community, providing health screening activities and educational materials.

Blue Cross and Blue Shield of Kansas City is proud to provide the best service in town to our customers every day. Rated among the top third of the nation's Blue Plans, the company maintains its high service standards by remaining focused on opportunities for improvement. This progressive outlook has led, in recent years, to the implementation of automated systems, online processes, and, ultimately, improved customer satisfaction.

In the big city and surrounding communities, people have come to rely on Blue Cross and Blue Shield of Kansas City for its strength and stability as well as its unwavering commitment to improve the overall health of the communities it serves. ♦

◀ *Blue Cross and Blue Shield of Kansas City is a major contributor to the PE4life program at Kansas City's Woodland Academy and Lincoln Middle School. PE4life is an innovative, national physical education and exercise program that makes fitness more fun, inviting, and interesting for kids of all ages. Blue Cross and Blue Shield of Kansas City is proud to sponsor PE4life in Kansas City.*

PHOTO BY MARIO MORGADO

Located in the renowned 18th & Vine Jazz District, the Negro Leagues Baseball Museum tells the story of a league that holds an important place in the sport's history. Encompassing ten thousand square feet, the museum houses a self-guided tour of manuscripts, photographs, film, uniforms, audio commentary, and more. Among the exhibits are twelve full-sized bronze sculptures of baseball greats, eleven of which have been inducted into the National Baseball Hall of Fame. Arranged in playing field formation, the bronzes include portrayals of notable names Satchel Paige, Rube Foster, and Buck Leonard. The museum also honors John Jordan "Buck" O'Neil, the twelfth bronze and a pivotal player in the Negro Leagues. O'Neil played for the Kansas City Monarchs and was a hometown hero, with batting titles and consistent averages of .345 and higher. After advancing to the status of all-star player and team manager, O'Neil worked his way into the white leagues as a scout for the Chicago Cubs, ultimately becoming the Major League's first African American coach with that organization. O'Neil's role in opening the door for African American players into the white leagues cannot be overstated; he is credited with signing such Hall of Famers as Lou Brock. In the late 1980s, O'Neil came back home to Kansas City to scout for the Royals organization, ultimately helping to form the museum and serving as its honorary chair until his passing in 2006. O'Neil was also prominently featured in Ken Burns's 1994 documentary, *Baseball*.

PHOTO BY DENNIS KEIM

GREATER KANSAS CITY: UNLIMITED POSSIBILITIES | 59

▲ *Child Life specialists offer fun methods of helping hospitalized children understand the medical care they are receiving—just one of the many ways in which Children's Mercy focuses on the unique needs of children and families.*

World-Class Pediatric Care: Where Kids Come First

When the levees burst in New Orleans following Hurricane Katrina in 2005, Randall L. O'Donnell, Ph.D., president and CEO of Children's Mercy Hospitals and Clinics, called the children's hospital in New Orleans to offer help. When he learned the hospital needed to evacuate, Dr. O'Donnell contacted U.S. Senator Kit Bond, who arranged for two military planes to rescue children from the flooding hospital. While other facilities accommodated two to four patients, Children's Mercy admitted twenty-four children and their families. Because of the excellent treatment they received at Children's Mercy, a number of those families have since made Kansas City their new home.

This event exemplifies the heroic efforts that have been part of the hospital's history since its beginnings. In 1897, two sisters—one a surgeon and the other a dentist—took in and cured a crippled five-year-old girl whose mother was trying to sell her on the street. As their generous practice grew, so did Children's Mercy Hospital. Today Children's Mercy includes two hospitals with a combined 314 beds, and more than fifty outpatient specialty clinics, providing state-of-the-art pediatric care. For more than a century, the hospital's mission has remained the same: to help as many sick

CHAPTER ONE: LIVING IN GREATER KANSAS CITY

children as possible through clinical care, education, and pediatric medical research.

The only hospital between St. Louis and Denver that caters solely to children, Children's Mercy provides approximately $12 million of charity care annually. Long-term patients receive tutoring in an in-house classroom or at their bedsides. Local patients and patients from around the world come to the hospital to receive the specialized services offered there—the surgeon renowned for his ability to operate on the walnut-sized heart of a newborn, critical care, the highest level of emergency and neonatal intensive care, and the full spectrum of pediatric medical and surgical inpatient and outpatient services. And Children's Mercy was the first hospital in Missouri or Kansas to receive the prestigious Magnet designation in recognition of excellence in patient care.

Medical and nursing students come to Children's Mercy, a teaching hospital, to learn from its highly qualified staff. They also learn the importance of attending to the emotional and psychological needs of sick children, an outstanding emphasis at Children's Mercy. The entire staff at Children's Mercy—with two Ronald McDonald House facilities available for patient families, playrooms on every floor, and beds for parents in every room—understands that easing a child's anxieties speeds healing.

"Parents are not considered visitors here," says Dr. O'Donnell. "Parents are an integral part of their child's recovery and part of our health-care team. Keeping life as normal as possible is a primary goal." The children here enjoy birthday parties, pet therapy, art and music therapy, entertainers, and Friday night carnivals. And when children are able to leave, the hospital provides home care whenever necessary. It's no wonder that the staff at Children's Mercy explains how far they will go to help a child by asking one simple question, "How high is the sky?" ♦

THE STAFF EXPLAINS HOW FAR THEY WILL GO TO HELP A CHILD.... "HOW HIGH IS THE SKY?"

PHOTO BY DENNIS KEIM

From the moment families step through the front door into this colorful and child-friendly lobby, they know that Children's Mercy is a place where children come first.

The Kansas City Zoo houses more than one thousand animals and approximately three hundred species. These ring-tailed lemurs (above) reside in the indoor Kidzone Discovery Barn, along with squirrel monkeys, meerkats, and macaws. Lemurs are small, sleepy primates that live in colonies and are found indigenously only on the island of Madagascar. In the zoo's most popular show, California sea lions, like Leon, shown (opposite) kissing his trainer, entertain spectators—with water routines, ball tricks, and general clowning—three times a day when weather permits. Even though the zoo has no special area designated as a petting zoo, children and adults can touch and feed many zoo inhabitants. This friendly fallow deer eagerly receives snacks from little Olivia Dumas (right), while Laura Mack (inset) enjoys a visit from a curious Australian rainbow larakeet that appears more interested in her cap than in the cup of nectar she is holding.

CHAPTER ONE: LIVING IN GREATER KANSAS CITY

Aquila: Delivering Power, Developing People

From its Kansas City headquarters, Aquila, Inc., oversees an operation of electric and natural gas distribution utilities that serve nearly a million customers in Colorado, Iowa, Kansas, Missouri, and Nebraska.

But this company's strength is derived from something far more powerful than its products. "Without question, the most important factor behind Aquila's success is our people," says Ivan Vancas, operating vice president of Aquila's Missouri electric operations. "Titles are really not important at Aquila; what really matters is ideas. We have many long-term employees, but we also have a number of new people who bring new ideas and energy to the team."

Working together, Aquila's people continually strive toward improved processes in order to ensure that energy is delivered in the safest, most efficient means possible.

That means affordable, reliable electricity, gas, and related services for both home and business. "Aquila has over twelve hundred employees in Missouri who come to work each day to serve our customers," says Vancas.

▲ *Eddy Halter (left) and Scott Hendricks helped restore a Kansas City home during Christmas in October. In 2006 Aquila participated in the annual event for the eighth consecutive year, when approximately 120 Aquila employees renovated five homes for needy families.*

CHAPTER ONE: LIVING IN GREATER KANSAS CITY

PHOTO BY DENNIS KEIM

PHOTO BY THOMAS S. ENGLAND

In the home, Aquila works to keep the power on for every one of its customers through energy savings and safety tips, financial aid, and initiatives like Aquila Cares, an energy assistance matching-funds program administered by the Mid America Assistance Coalition and the United Way of Greater St. Joseph.

In the office, Aquila provides competitively priced electricity, natural gas, and other related energy products that help business and industry thrive.

Aquila also fosters business growth through programs that emphasize economic development. Professionals in this segment of the company partner with communities to understand their needs in an effort to attract new business and industry and support the retention and expansion of the local business base.

"We have very strong community relations and economic development teams who work with communities at the grassroots

(continued on page 66)

WORKING TOGETHER, AQUILA'S PEOPLE CONTINUALLY STRIVE TO ENSURE THAT ENERGY IS DELIVERED IN THE SAFEST, MOST EFFICIENT MEANS POSSIBLE.

◀ *Aquila apprentices Luke Johnson (left) and Mike Schoon install a transformer during construction at North Kansas City High School. The high school has an energy-efficient geothermal heating and cooling system, which saves the district up to 50 percent on its heating and cooling costs.*

AQUILA

(continued from page 65)

level each and every day," explains Vancas, expanding on this group's activities. "We offer the LocationOne information system that allows communities to market facilities to businesses and industries across the region and the country. We facilitate strategic planning workshops with community leaders to identify strengths and areas of improvement and then work with them to make those plans come alive. We provide financial support to city, county, and regional economic development professionals to further their education and increase their skills. And we offer support with writing grants and sponsoring studies."

Beyond growing business, Aquila and its people place equal importance on helping communities thrive. That's why you'll find the people of Aquila working hard for organizations that try to make the world a better place. "Aquila is a public service company, and we think civic involvement and community service are fundamental to what we do," says Vancas. "Aquila's philosophy is that, if our communities grow and prosper, then Aquila will grow and prosper as well. Our employees volunteer for a wide variety of activities because they care about the communities where they live and work. This makes them, and our company, special."

Philanthropic efforts are targeted at youth development, education, literacy, environment, and human services as areas that provide the basic qualities of vibrant neighborhoods. Among the 320 entities that Aquila supports are the United Way, National Night Out, Christmas in October, Caring for Kids, school supply drives, and low-income weatherization.

From reliable delivery of electricity and gas, to people taking time to make a difference, Aquila is a utility that works hard to energize the communities it serves. ♦

PHOTO BY THOMAS S. ENGLAND

▲ *State-of-the-art office design, extremely efficient use of energy, and advanced communications systems combine at Aquila's 20 West Ninth headquarters in Kansas City, Missouri. The renovated former New York Life Building touts a high-efficiency electric chiller to help lower cooling costs.*

PHOTO BY THOMAS S. ENGLAND

◄ *Jim Dixon (left) stands by his heat pump installed by Climate Control Heating and Cooling, an Aquila PowerTech dealer. David Dennis, president, stands nearby. Dixon chose a heat pump to save money and make his home more energy efficient.*

CHAPTER ONE: LIVING IN GREATER KANSAS CITY

Since its opening in October 2006, the Center for Advanced Heart Care houses one of the most patient-driven, quality heart programs in the region. Located at the University of Kansas Hospital, the center comprises seven floors, housing over one hundred beds for inpatient and outpatient care. In addition to offering some of the latest techniques in heart care, the center borrowed concepts from the hospitality industry to create an environment that promotes healing. Among the center's amenities are overnight accommodations for caregivers, wireless computer access, a learning and resource center, specially designated rehabilitation areas, and decentralized nursing stations that place care within easy reach of patients. The center also includes a newly expanded emergency department and was built with expansion in mind.

PHOTO BY DENNIS KEIM

PHOTO BY DENNIS KEIM

Educating the next generation of top physicians remains a key part of the mission of the University of Kansas Hospital. Here, Dr. Gary Gronseth reviews a patient record with his residents. Embarking on a new direction in 1998 to become independent of state funding, the hospital has successfully enhanced its ability to fulfill its mission as the region's premier academic medical resource for health-care providers of today and tomorrow.

The Nelson-Atkins Museum of Art draws visitors like Brook Edgington Griffin, Sierra Buffington (above), and Deborah Gray (right) to view the encyclopedic collection of more than thirty-four thousand works of art. The pieces represent all cultures and times and span five thousand years. The local spark of creativity that made the museum possible came from the vision and generosity of William Rockhill Nelson, founder of what is now the *Kansas City Star*, and schoolteacher Mary McAfee Atkins.

CHAPTER ONE: LIVING IN GREATER KANSAS CITY

Exploring the collections housed at the Toy and Miniature Museum of Kansas City is like going on a treasure hunt, but in four different centuries. Housing one of the country's largest and most comprehensive collections of antique toys, folk art, and miniatures, the thirty-three-thousand-square-foot museum features everything from seventeenth-century dolls to modern-day marbles. It is also well known for the depth and breadth of its mini–decorative arts collection—from period furniture to paintings and musical instruments—as well as an extensive collection of micro-miniatures. Created entirely by hand with the aid of a microscope and special tools, these micro-miniatures are an art unto themselves and even include fleas dressed up in various costumes, which is no small feat.

PHOTO BY SCOTT INDERMAUR

GREATER KANSAS CITY: UNLIMITED POSSIBILITIES | 69

Johnson County Government: Providing Key Services for a Thriving Community

Nestled in the southwestern quadrant of the metropolitan area, Johnson County is the economic engine driving success in the metro region and the Sunflower State. With more than a half-million residents, Johnson County is the most populous county in Kansas, enjoying an average increase of about ten thousand new residents each year. It's recognized throughout the nation as a premiere community—a place where people want to live, work, and raise a family.

With a national reputation for excellence, Johnson County exhibits all the hallmarks of a great community: a thriving and growing business sector offering economic opportunities and jobs, open spaces, cultural amenities, a well-designed and maintained infrastructure grid and transportation corridors, and distinctive neighborhoods that welcome people, whether they've lived here for fifty years or just arrived.

Despite its strong, steady growth rate, roughly 55 percent of Johnson County's

▲ *The Johnson County Park and Recreation District offers plenty of opportunities to enjoy some cooling-off fun, including a public swimming pool at the Roeland Park Aquatic Center, along with outdoor facilities at the Heritage Park Marina and Kill Creek Park Beach and Marina, both in Olathe, and the Shawnee Mission Park Beach and Marina, Shawnee.*

477 square miles are divided among nineteen different cities, leaving nearly 45 percent of land available for future growth and development. That growth is important to a robust and diverse local economy, a place that is home to the operations of half of all Fortune 100 companies and one-third of all Fortune 500 companies. Businesses enjoy a supportive environment in Johnson County, due in large part to its central geographic location and ease of accessibility from all parts of the United States. Additionally, residents enjoy a low cost of living and a low crime rate, while having access to award-winning schools, libraries, and parks—all of which combine to make Johnson County a highly desirable place to do business and raise a family.

Johnson County Government is headquartered in the county seat of Olathe. Legislative and policy-making powers are vested in a board of county commissioners that includes six members elected by district and a chairman elected at-large from the entire community to a four-year term as the county's chief elected official. The County Government's services are provided by thirty-nine agencies and departments under the direction of a county manager who serves as chief administrative officer. Johnson County is a "full service" local government, providing a range of services from aging to zoning. In addition to the traditional county services—including law enforcement and public safety, the collection and disbursement of taxes, the administration of elections and voter registration, public works, and health and human services—Johnson County Government also owns and operates two airports (the third- and fifth-busiest in Kansas), coordinates homeland security and emergency management services, exercises land use and zoning regulatory powers, operates a public transit system, and manages a wastewater system serving more than 380,000 customers.

"Johnson County is a community on the move—a community that's building and moving forward and making progress every day," said Annabeth Surbaugh, chairman of the Johnson County Board of Commissioners. "At the heart of our success story is an unwavering commitment to building a better future for tomorrow's generations, and that guides our decisions today in sustaining Johnson County as a *Community of Choice.*" ♦

PHOTO BY ERIC FRANCIS

AT THE HEART OF ITS SUCCESS IS AN UNWAVERING COMMITMENT TO BUILDING A BETTER FUTURE FOR TOMORROW'S GENERATIONS AS THEIR *COMMUNITY OF CHOICE.*

◀ *The Children of the Trails statue and fountain were dedicated in September 2000 and are located on the court square between the Johnson County Courthouse and Johnson County Administration Building in downtown Olathe. The statue was created by Kwan Wu.*

The Ewing and Muriel Kauffman Memorial Garden is an exquisite area of lush and colorful beds of annual and perennial flowers, the soothing sound of fountains, and original bronze sculptures. The garden is an inviting outdoor feature of Kauffman Legacy Park, home to the renowned Ewing Marion Kauffman Foundation, the Kauffman Foundation Conference Center, offices of the Missouri Department of Conservation, and the Discovery Center.

With the support of the Ewing Marion Kauffman Foundation, FIRST Robotics has encouraged a record number of Kansas City-area high school students to apply math and science principles to design, assemble, and test robots capable of performing specific tasks. The goal of the FIRST project, which stands for "For Inspiration and Recognition of Science and Technology," is to help young people recognize their abilities and inspire excitement in problem-solving using science, technology, and engineering. Teams have six weeks to complete their robots, using only their own inventiveness. Teams are guided by volunteer mentors from the area's engineering, academic, and related professions.

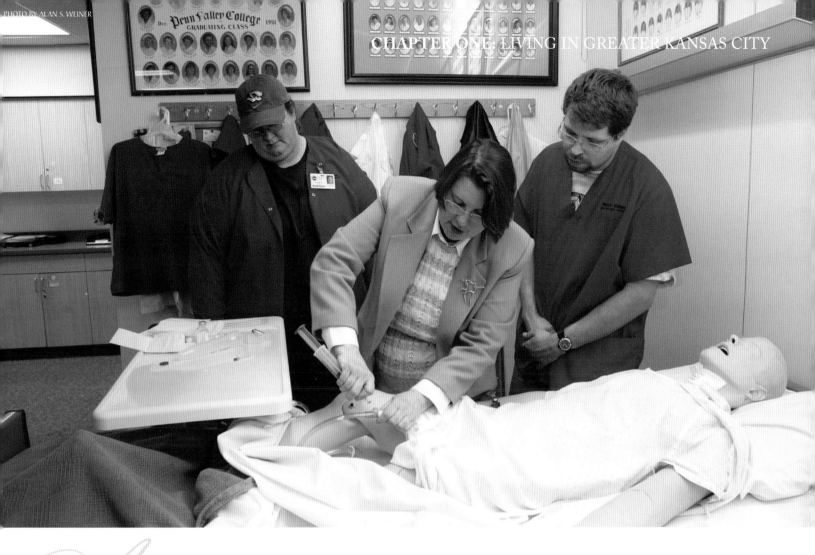

A nursing instructor at the Metropolitan Community College (MCC) demonstrates an irrigation technique to students. MCC is home to more than a dozen health-care programs, making it a vital source for these in-demand careers. The programs are also in demand on campus, thanks to a combination of quality instruction and affordability. Most are located at the MCC–Penn Valley campus in midtown Kansas City. Programs include both practical and professional nursing, dental assisting, physical therapist assistant, health information technology, coding specialist, medical transcriptionist, radiologic technology, occupational therapy assistant, paramedic, respiratory care, surgical technology, and more.

One of the premier automotive technology programs in the country is located at the Metropolitan Community College (MCC) Longview campus in Lee's Summit. The program partners with Ford and GM, which provide the latest equipment and cars. The menu of more than eighty career programs at MCC campuses includes an amazing variety of careers that can be learned in two years or less. These include HVAC, horticulture, computer-aided drafting and design, graphic design, networking, game programming, geographical information systems, child growth and development, more than a dozen health-care careers, and an academy for peace officer or firefighter training.

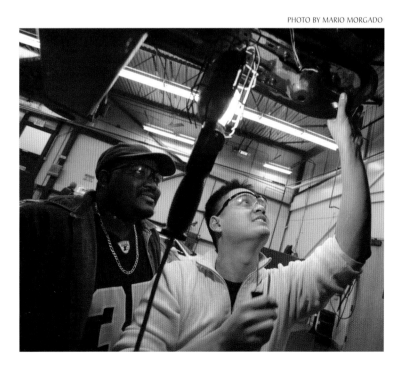

The Greater Kansas City Community Foundation: Improving the Region's Quality of Life

"One of the best things the Greater Kansas City Community Foundation can do is to help a donor deepen his or her charitable passion," said Laura McKnight, president and CEO. "We always ask, 'What is important to you? What would you like to see improved in Kansas City to make it an even better place?'"

The founders of the Greater Kansas City Community Foundation passed a hat in 1978, collecting $219.13, believing that philanthropy is the opportunity and responsibility of everyone, not just a few. Since then, the foundation has grown to over $1 billion in assets, spread among more than eighteen hundred funds dedicated to the causes that are important to the individuals, families, and businesses that established them.

Now one of the ten largest community foundations in the country, and with over $1 billion in grants distributed since inception, the Greater Kansas City Community Foundation works side by side with Kansas City's donors to fulfill community dreams through the power of giving.

"This is the community's foundation," McKnight said. "We take seriously our mission to improve the quality of life in Kansas City by involving everyone—all races and ages and our entire metropolitan area, north, south, east, and west—in the joy of making a difference." ♦

▲ *The Greater Kansas City Community Foundation takes seriously its mission to improve the region's quality of life, because Kansas City is a great place to live.*

"WE TAKE SERIOUSLY OUR MISSION TO IMPROVE THE QUALITY OF LIFE IN KANSAS CITY."

CHAPTER ONE: LIVING IN GREATER KANSAS CITY

There are many reasons to study the martial arts, and self-defense is just one of them. At the American Taekwondo Association's Karate for Kids program in Kansas City, children between four and fourteen develop the discipline and leadership skills that will serve them in other aspects of their lives. They also get a chance to participate in yearly ATA-sanctioned events like this one. A recent tournament event featured teams from thirteen states, including Missouri. In addition to its Karate for Kids program, the Kansas City ATA school hosts a Tiny Tigers program for children ages four to six that teaches motor skills, develops the ability to take direction, and builds a strong foundation in character qualities such as courtesy, respect, and discipline. Once a student reaches fourteen years of age, he or she can enter the school's adult classes, which offer advanced training in the martial arts as well as training certification programs for potential instructors.

GREATER KANSAS CITY: UNLIMITED POSSIBILITIES | 75

KCMO Water Services Department Maintains and Provides a Vital Life Source

During the earliest days of Kansas City, most residents drew water from cisterns and wells, but waterborne illness was rampant, and fire protection lacking. In 1874, the city commissioners selected the National Water Works Company of New York to build and operate the city's first waterworks. With twelve miles of mains and a capacity of 5 million gallons per day, operations began in 1875, giving the city "good" water from the Kaw River. In 1887, the water supply was switched to the Missouri River.

In 1895, the city bought out the National Water Works Company, forming what is today the KCMO Water Services Department. The department expanded the city's water system and in 1920 took on the largest civic project to date: construction of the Kansas City Water Treatment Plant and a three-mile-long tunnel to deliver the water to residents south of the river. By 1930, the Kansas City Water Treatment Plant was completed and able to provide residents with up to 100 million gallons of water per day.

Today, the department produces an average of 140 million gallons per day, transforming the mighty Missouri River into safe, clean water that meets or exceeds all state and federal guidelines, and which in fact has been recognized as the best water in the country. In addition, the department is responsible for the city's wastewater system and the city's stormwater management program.

▲ *The Kansas City Water Treatment Plant was completed in 1930 and today provides the city with up to 240 million gallons of water per day. Construction began on the plant in 1924.*

"Our goal is to be the finest water supply, wastewater treatment, and stormwater management utility while maintaining excellent customer service and water quality at a reasonable price in an environmentally friendly manner," says director Frank Pogge.

The department is continually upgrading and maintaining the system, which includes more than twenty-three hundred miles of water mains, twenty-seven hundred miles of sewer lines, more than thirty thousand storm drain inlets, and twenty-two thousand fire hydrants.

In addition, the department is currently working on a comprehensive program, Wet Weather Solutions, to address sanitary sewer and stormwater issues. The program includes a long-range plan to manage wet weather flow in both the combined and separate sewers within the city, as required by the U.S. Environmental Protection Agency and the Missouri Department of Natural Resources. The Combined Sewer Overflow Control Plan addresses sewer overflows from the combined sewer system. The Sanitary Sewer Overflow Control Plan addresses overflows from the city's separate sanitary sewer system. The program will possibly be the largest infrastructure program in the city's history.

"The department has been serving the city for more than one hundred years and realizes that the present and future depend on the department's commitment to providing safe, quality drinking water," Pogge says. "At the same time, we strive to treat wastewater in an environmentally friendly manner, protect the city from flooding, and protect stormwater quality." ♦

"WE STRIVE TO TREAT WASTEWATER IN AN ENVIRONMENTALLY FRIENDLY MANNER, PROTECT THE CITY FROM FLOODING, AND PROTECT STORMWATER QUALITY."

◀ The sun may be shining today, but when the weather turns rainy, the city's Wet Weather Program is there to keep the water moving. The program includes two separate but coordinated planning processes: the Kansas City Overflow Control Program and KC-One Stormwater Management Plan. Together, these two plans address flooding, sewer backups, water quality, and wastewater overflows.

*S*ometimes it take years for technology to catch up with the imagination of an artist. That was surely the case with the Steeple of Light conceived by Frank Lloyd Wright for Community Christian's Church of the Future. The notes on his architectural renderings called for "searchlights piercing the perforated masonry roof . . . sky-beams." However, that was in 1940, and the technology did not exist to fulfill the dream. In 1990, Dale Eldred ran some initial tests with various available lights, but again, the technology wasn't ready. In 1994, a community group approached Eldred's widow and artistic collaborator about revisiting her late husband's vision. A lighting firm in Tennessee was contracted to make the lights, and on December 15, 1994, the Steeple of Light finally came to life. Each of the four lights weighs approximately three hundred pounds, and produces 1.2 billion candlepower of illumination, reaching several miles into the stratosphere and visible from ten miles away.

CHAPTER ONE: LIVING IN GREATER KANSAS CITY

Since 1917, the Jewish Community Center of Greater Kansas City has grown and adapted to serve the changing needs of the city's Jewish—as well as secular—community. One special observance is Hanukkah, which is also referred to as the Festival of Lights or the Festival of Dedication. The menorah's candles are lit each night of the festival, as Sammy Chervitz is doing. Friends and family joining him are (left to right) Beatrice Fine holding her son, Benjamin, Sammy's sister Sophie, baby Nathaniel, his mother Vicki, and his friend Shifra Dimbert.

CHAPTER ONE: LIVING IN GREATER KANSAS CITY

a. Cathedral of the Immaculate Conception

b. Community of Christ Temple

c. Rime Buddhist Center in Kansas City, MO

d. Rime Buddhist Center in Kansas City, MO

e. Islamic Center of Greater Kansas City

f. Kehilath Israel Synagogue

g. First Presbyterian Church in Leavenworth, KS

h. Cathedral of the Immaculate Conception

i. White Church Christian Church

Throughout the Greater Kansas City region, there are as many lovely places of worship in as many different architectural styles as there are religious denominations. Most of them have quite an interesting history. The Cathedral of the Immaculate Conception, whose tower has been a feature of the Kansas City skyline since 1884, is the mother church of the diocese of Kansas City. It has the oldest worshiping congregation in Kansas City and dates back to a log cabin church in 1834. First built by the Delaware Indians in 1832, the White Church Christian Church is listed in the Kansas City Registry as the only church in the state with continued worship on the same site since 1832. The First Presbyterian Church of Leavenworth, Kansas, can trace its history to an outdoor church service alongside the Missouri River in October 1854. Formerly known as the Reorganized Church of Jesus Christ of Latter Day Saints, the Community of Christ Temple in Independence, Missouri, traces its roots back to 1860. Built in response to a revelation presented at the community's 1984 World Conference, the temple was designed by Japanese American architect Gyo Obata to mimic the spiral shell of the Nautilus, a theme that is continued up the temple's three-hundred-foot-high stainless-steel spire.

GREATER KANSAS CITY: UNLIMITED POSSIBILITIES | 81

◀ *The three-story atrium lobby is one of the most striking features of Truman Medical Centers' Lakewood facility. When planning the building, staff and architects focused on eliminating the clinical look of a hospital. With warm colors, curves, and natural light, woods, and fabrics, the hospital's comforting environment also meets or exceeds regulatory requirements and was designed with a very efficient work flow in mind.*

PHOTO BY DENNIS KEIM

Truman Medical Centers: Quality, Compassion, Technology, and Innovation

Truman Medical Centers (TMC) offers health-care services to any person in need, an approach to service that earns TMC the description of "primary safety-net provider" of inpatient and outpatient community care. TMC Hospital Hill, recently named one of the nation's top academic medical centers, specializes in trauma services, diabetes care, and bariatrics. TMC Behavioral Health operates an inpatient and several outpatient facilities. TMC Lakewood, with a new 175,000-square-foot addition, is home to one of the country's largest family residency programs.

The Lakewood addition was designed with the goal of being as hospitable as a medical building can be for patients and staff. Constructed using LEED (Leadership in Energy and Environmental Design) standards, Lakewood's interior, even the ICU, is awash with natural light and warm colors. "The TMC staff remains committed to offering not only state-of-the-art technology, but also state-of-the-art care," states Dolores Sabia, business development and ambulatory care officer.

The TMC system has been a part of Jackson County's health-care services since 1908, but it has embraced future technologies faster than many of its contemporary counterparts. Named as a Most Wired Hospital by Hospitals and Health Networks, TMC is one of the only area medical groups with PACS (a photo archiving computer system). PACS enables physicians to immediately access patient X-rays from remote locations, even from the operating room, meaning that TMC is poised to safely care for patients well into the future. ♦

TMC IS POISED TO SAFELY CARE FOR PATIENTS WELL INTO THE FUTURE.

CHAPTER ONE: LIVING IN GREATER KANSAS CITY

Originally built as a school, Webster House provided a public education for thousands of Kansas City children from 1885 until 1932. Today, the stately Romanesque building has been given new life as an antique and interior design establishment. As the oldest extant schoolhouse in Kansas City, it has also received numerous awards for its restoration and design and is listed on the National Register of Historic Places. Showcasing one of the largest collections of antiques in the Midwest, each room in Webster House is filled with surprises, including exceptional examples of eighteenth- and nineteenth-century European, American, and Asian fine arts, furniture, and accessories. Modern-day gifts and accessories are also available, along with a full-service interior design department to help clients pull it all together. Another surprise is the on-site Webster's Restaurant, a destination in and of itself for elegant lunches featuring critically acclaimed New American cuisine.

GREATER KANSAS CITY: UNLIMITED POSSIBILITIES | 83

Mark One Electric Company Inc.: The Mark of Electrical Excellence

Innovative thinking coupled with the ability to get the job done right the first time has made Mark One Electric Company Inc. one of the region's top electrical contractors. Specializing in all facets of electrical and utility construction, Mark One offers its clients a wide range of services, including design/build construction, estimating, preconstruction coordination, project management, and architectural design.

Mark One's founder, Carl "Red" Privitera Sr., led the company from 1974 until 1994. During that time, his leadership and vision established the company as one of the finest industrial and commercial electrical contractors in the community. His client list included some of the biggest names in industry: General Motors, the Ford Motor Company, SWB, and AT&T.

Following Red's retirement in 1994, Rosana Privitera Biondo became the new president of Mark One Electric, with her brothers, Joseph, Carl, and Anthony Privitera, serving as vice presidents to provide the ultimate leadership team in the industry. Today, Mark One continues to thrive, completing the new IRS Service Center, the Nebraska Furniture Mart expansion, the Kansas City International Airport consolidated car rental facility,

▲ *Not only does Mark One excel at design/build projects, it also provides the services needed to maintain those projects. Whether a client needs to replace a light switch or an entire transformer, Mark One's team of certified electricians are available 24/7 to help. Here, they have been called in to inspect the wiring on a piece of equipment for the Shoal Creek Police Station.*

▶ *Mark One electricians pull cable during construction of the new Kansas City Police Academy. As part of its specialty services, Mark One also employs a team of certified installers capable of working with high-tech communications, fiber optics, fire, security, and access control systems.*

CHAPTER ONE: LIVING IN GREATER KANSAS CITY

INNOVATIVE THINKING COUPLED WITH THE ABILITY TO GET THE JOB DONE RIGHT THE FIRST TIME HAS MADE MARK ONE ELECTRIC CO. INC. ONE OF THE REGION'S TOP ELECTRICAL CONTRACTORS.

and the *Kansas City Star* production plant.

The Mark One management team's understanding of owner, architects', and engineers' difficulties comes from their own experiences of managing and developing the properties that their parents, Carl and Josephine, own. This experience and the lessons learned allow Mark One to guide clients through the various phases of development, from site selection all the way through construction.

Certainly, it is an exciting time to work at Mark One, with more new construction under way in the Kansas City area than at any time since the Pendergast era. Mark One is encouraged that work for all the construction crafts will be in demand in the area for some time to come. ♦

▲ Building the city's new IRS Service Center was part new construction, part renovation of the former main postal facility. As a result, one aspect of Mark One's work as the project's primary design/build electrical contractor was to relocate all of the existing underground electrical and communication utilities and construct a new primary power feed. The new campus includes a 1.4-million-square-foot processing facility, a one-hundred-thousand-square-foot renovation of the Railway Express Building at Union Station for the United States Postal Service (located across the street), and over 1.2 million square feet of above- and below-ground parking.

CHAPTER ONE: LIVING IN GREATER KANSAS CITY

PHOTO BY DENNIS KEIM

Jazz storytelling at the American Jazz Museum in Kansas City, Missouri, involves children in interactive lessons (including games, musical instruments, and stories) focused on jazz, an art form that thrived in the city in the early 1900s. Kansas City is where jazz greats like Count Basie and Charlie Parker got their start, and the music remains an essential part of the culture today. In free-to-the-public sessions, vocalist Lisa Henry, professional storyteller Brother John, and a group of jazz musicians teach rhythm, movement, and the history of jazz and its legends through singing, dancing, and scatting. Anywhere from 50 to 150 children attend each twice-a-month program. Participants especially love learning and performing the song "Where You At? The Jazz Museum!" The lessons end in a jazz mambo line, and even though the program was designed with children in mind, more often than not, parents and even elderly groups get involved in the action. "It's contagious because they're having so much fun," said Dennis Winslett, educational specialist.

GREATER KANSAS CITY: UNLIMITED POSSIBILITIES | 87

Metropolitan Community College: Opening the Doors to Opportunity

At Metropolitan Community College, learning is a process of discovery. With a mission to prepare students, serve communities, and create opportunities, this publicly supported two-year college not only helps students discover their future, it encourages cultural and economic growth throughout Greater Kansas City.

The roots of Metropolitan Community College go back to 1915, when the Kansas City Polytechnic Institute was founded at Eleventh and Locust streets. In 1919, the institute became the Junior College of Kansas City and was the first two-year college in the United States to award an associate's degree. With five locations—Blue River, Business & Technology, Longview, Maple Woods, and Penn Valley—MCC has grown into Kansas City's largest institution of higher learning. Currently, the college enrolls over forty-three thousand students per year and employs more than three thousand people.

As a fully accredited institution, MCC is committed to eliminating barriers to the pursuit of higher learning. Continually striving to reach the highest level of quality in its processes, programs, and services, MCC offers its students an education that equals, if not exceeds, that of a four-year

▲ *More than 70 percent of Metropolitan Community College students plan to transfer to a four-year institution. For two years, they work on general education requirements and earn a degree while enjoying small personal classes, excellent instructors, and affordable tuition.*

institution. As a result, nearly half of its students choose to complete their first two years of bachelor's studies at MCC before transferring to a four-year college.

Not only does MCC attract top-notch instructors, many of them with real-world experience, it allows them the luxury of teaching in small classes. With the average class containing about eighteen students—nearly a quarter of the size of typical freshman classes at a four-year university—students receive personal attention on a daily basis. Student support services are also extensive and range from financial aid and tutoring to career advisement and placement to veterans' benefits and single-parent assistance. Finally, its tuition is less than one-third that of a public university.

In addition to more than a dozen associate's degree programs covering the arts, sciences, and engineering, Metropolitan Community College also offers over eighty two-year career programs. Defined by quality instruction and hands-on experience, these programs attract students interested in such hot career fields as health care, networking, computer technical support, multimedia technology, accounting, and public safety. From arts to engineering, hospitality to human services, manufacturing to veterinary technology, the possibilities for a new career or expanded job skills are virtually endless.

Of course, it's not all about the classes. Extracurricular opportunities are nearly as diverse and extensive as those found at a four-year university and include intramural athletics, honor societies, dozens of specialized clubs, theatrical production, film festivals, a nationally competitive debate team, and much more.

The environment at MCC is one that encourages success, academic integrity, diversity, and barrier-free inquiry. Its relationship with the outside world is also defined by caring commitment. In preparing students for their future, MCC positively impacts the future of Greater Kansas City. Its partnerships with area businesses and economic development groups help train workers, retain jobs, and bring new job opportunities to the entire area. ♦

WITH A MISSION TO PREPARE STUDENTS, SERVE COMMUNITIES, AND CREATE OPPORTUNITIES, MCC ENCOURAGES CULTURAL AND ECONOMIC GROWTH THROUGHOUT GREATER KANSAS CITY.

◀ Career training is a major part of the educational opportunities that Metropolitan Community College provides. More than eighty career programs are offered, including technical skills training. Here, manufacturing technology students get hands-on training that prepares them for immediate employment.

Once a stopover for wagons traveling the Santa Fe Trail, Hyde Park remains one of Kansas City's most beloved neighborhoods. Homes built in the area at the turn of the century were considered to be some of the most distinctive addresses in town and offered a blend of architectural styles ranging from Victorian to Prairie. Today, the area retains its charm, and a portion of Hyde Park is listed on the local and national historic registers.

CHAPTER ONE: LIVING IN GREATER KANSAS CITY

Kansas City Regional Association of REALTORS®: The Voice of Real Estate

Most people turn to professionals when buying and selling property, and with good reason. REALTORS® have the tools and talent to make the transactions as efficient and as effective as possible for the consumer. And in Kansas City, the Kansas City Regional Association of REALTORS® (KCRAR) has been the voice of real estate for more than a century.

"As a trade association of more than ten thousand real estate professionals, KCRAR's ultimate goal is to promote property ownership and the reality of the American Dream," says Kathy Copeland, president, KCRAR board of directors. The not-for-profit association, formed in 2001 by the merger of the Johnson County Board of REALTORS® and the Metropolitan Kansas City Board of REALTORS®, is one of the few of its kind with reach on both sides of a state line. Its membership represents all areas of real estate, including residential sales, commercial sales and leasing, development, property management, and businesses affiliated with real estate. If a real estate transaction occurs in Kansas City, Peculiar, Leavenworth, Gardner, Liberty, or anywhere in between, chances are that a KCRAR member is involved.

In addition to holding the Heartland Multiple Listing Service (HMLS)—a complete and accurate source of real estate

▲ *Residential Realtor members of KCRAR use the Supra electronic lockbox system. These secure lockboxes are easy to use, and the variety of electronic key devices can consolidate many tools of the trade into one.*

CHAPTER ONE: LIVING IN GREATER KANSAS CITY

information in the Greater Kansas City metropolitan area—KCRAR provides a number of other services for its members.

"As our real estate industry evolves, it is imperative for our members to keep up to date on all aspects of the industry through our continuing education program. These classes assure that members have the opportunity to keep abreast of changes in real estate to provide the highest and most professional service to the consumer," Copeland says. KCRAR offers continuing education classes, with subjects ranging from trends in the industry to relevant laws and regulations. Education programs also cover specific professional skills, including opportunities for members to train for and acquire esteemed industry designations and certifications.

In addition, KCRAR operates the REALTOR® Store, where members can purchase maps, signs, sign riders, technology products, forms, contracts, and other supplies necessary to enhance their business and effectively market their services. The association understands the importance of technology in today's world and, as such, helps members learn about and utilize such tools. "Both HMLS and KCRAR offer the latest in technical seminars and classes created to keep our members on the cutting edge of technology. With more than 80 percent of buyers searching for homes first on the Internet, it is critical for the real estate community to be able to communicate and provide information in the format the consumer desires."

Another vital role of KCRAR is advocacy. The association is involved in government activities and legislation that affect the real estate industry. "Our members are active on a local, state, and national level with our legislators, assuring that residential and commercial property owners' rights are considered and protected. Working hand-in-hand with government partners provides grassroots information on the importance of property ownership for all individuals. As an association, we're here for our members," adds Copeland. "And in so doing, we help build communities." ♦

"AS AN ASSOCIATION, WE'RE HERE FOR OUR MEMBERS, AND IN SO DOING, WE HELP BUILD COMMUNITIES."

◀ KCRAR members are more than Realtors. They are savvy businesspeople who are dedicated to helping clients find the optimal location and space for their businesses. The Crown Center area near downtown is just one of the many locations to consider.

CHAPTER ONE: LIVING IN GREATER KANSAS CITY

In 1917, George E. Powell Sr. left the family farm for a successful career in Kansas City, Missouri, but he never lost his love for the land. Powell eventually purchased 915 acres that he generously shared with the community, transforming it into a botanical garden in 1988. Powell Gardens, east of Kansas City, is available to the public for a scenic stroll among the blossoms or as a heavenly site for a wedding. From the viewing pier, visitors can clearly see the surrounding two-acre Island Garden's waterfalls and water plants, as well as the Marjorie Powell Allen Chapel. The chapel's all-glass front faces the water's edge, allowing a relaxing view of the lake, stippled with a marvel of light and shadows courtesy of the wood-and-window architecture. An interactive fountain, nature trail, conservatory, an annual Festival of Butterflies, and a free seasonal trolley make a visit to Powell Gardens a kaleidoscopic event.

Johnson County Community College: Quality Education for a Growing Community

In 1969, when the residents of Johnson County voted to purchase more than two hundred acres for a community college site, they envisioned a facility that served a broader mission than that of a four-year college. Today, Johnson County Community College (JCCC) fulfills that vision by providing quality undergraduate courses, developmental education, career training, and workforce development for a diverse community. With more than thirty-five thousand students enrolled each semester, JCCC offers a full range of transferable college credits, telecourses, online classes, and more than fifty career and certificate programs.

"You can get your first two years here, receive education for a new job, or take classes to keep your skills current," explained Julie Haas, director of college information. "We also have one hundred transfer agreements with other colleges and universities." JCCC's many accolades include the Kansas Excellence Award and a spotlight in *Rolling Stone* magazine as one of the country's most respected community colleges. Here, quality of instruction is the foremost goal, bolstered by small class sizes and affordable tuition, but the quality of campus life is also a strong draw. JCCC provides students access to excellent indoor/outdoor athletic facilities, a performing arts center with several theatres, and the Nerman Museum of Contemporary Art. The school's positive economic impact on the community totals $182 million annually, making JCCC not only a support to its students but to the community at large. ♦

▲ *The Student Center is the heart of Johnson County Community College. Known for high-quality classes and teaching, the college serves more than thirty-five thousand credit and continuing education students each semester.*

QUALITY OF INSTRUCTION IS THE FOREMOST GOAL, BOLSTERED BY SMALL CLASS SIZES AND AFFORDABLE TUITION.

CHAPTER ONE: LIVING IN GREATER KANSAS CITY

Independence, Missouri, is designated as a Preserve America Community, an honor recognizing community efforts to preserve historical sites and heritage. The city especially honors Harry S. Truman, the thirty-third U.S. president, who was born in Independence in 1884. This statue stands in Independence Square downtown, and is but one of the historic monuments dedicated to the great man who guided America through the close of World War II. The Trumans lived in this Victorian residence before and after his presidency, but Truman never actually owned the house. It was Mrs. Truman's family home. Guided tours are available to the public, including a glimpse into the Truman study, which contains 1,013 books. In 1953, Truman stated, "I look forward to a grand time doing what I please and going where I please." He never fully realized that retirement dream, as he continued to serve his country in a variety of capacities until his death in 1972.

GREATER KANSAS CITY: UNLIMITED POSSIBILITIES | 97

What little girl doesn't dream at some point of being a dancer? At Diane's School of Dance, five- and six-year-old girls get to live that dream in classes designed to teach them the basics of tap, ballet, and acrobatics. Owned and operated by Diane Henderson for over thirty years, Diane's School of Dance does more than just teach technique. While many dance studios are strictly competition-oriented, Diane's is geared more toward making dance a fun after-school activity where children can also learn basic manners and social skills. And everyone gets to dance. The school puts on yearly Christmas performances and an annual revue at the Kansas City Music Hall. In addition to teaching kids aged three to eighteen, Diane's also offers tap and jazz classes for adults.

PHOTO BY DENNIS KEIM

PHOTO BY DENNIS KEIM

CHAPTER ONE: LIVING IN GREATER KANSAS CITY

In 1988, a group of Kansas City, Missouri, citizens persuaded Mrs. Hema Sharma to open a local school for teaching the classical dances of India. Nritya, the name of the school, is also the term for a type of dance that portrays mood with facial expressions, hand gestures, and specific leg and feet positions. After years of diligently promoting the culture of India, Nritya's performances have become respected and sought after throughout the Midwest. Sharma, who since the age of twelve has danced more than five hundred times for audiences in the United States and India, is now the artistic director of the school. She and her dance artists/instructors especially enjoy teaching the younger generations about Indian art and music.

Park University Serves Students Worldwide

Park University, founded in 1875, is unique in many ways. It has forty-three campuses nationwide and online, including four campuses in the Kansas City area: Downtown, Independence, Parkville, and Lexington. More than 60 percent of the enrollment of approximately twenty-five thousand students is active-duty military personnel, military dependents, retired military, or associated with the Department of Defense.

Although it has a long and impressive history, Park is also on the cutting edge of technology. In 2006, the school was ranked second by U.S. News & World Report for its online enrollments. The flexibility of graduate and undergraduate options is also impressive. The Accelerated Degree Program, for instance, is built on eight-week terms, with classes meeting once a week for four and a half hours, to give working adults a convenient way to finish a degree. Many professionals living and working downtown take advantage of this flexible program.

"We've had a downtown presence for more than thirty years," said Rita Weighill, associate vice president for communication. "Park was one of the earliest local universities to offer weekend, evening, and online classes for adult learners, and it continues to

▲ *Founded in 1875, Park University has a long history of educating students for success in the global marketplace. More than twenty-one thousand students annually take advantage of Park's undergraduate and graduate programs that allow for meaningful academic discussions with the faculty.*

CHAPTER ONE: LIVING IN GREATER KANSAS CITY

redefine those areas. Our enrollment at the Downtown campus continues to grow as we add new graduate and undergraduate degree programs. We offer master's programs in public affairs, business administration, health-care leadership, education, teaching, and communication and leadership. In addition, we have added two new graduate certificate programs: health-care advocacy and computer and network security (CNS)."

"Working in government, especially with my antiterrorism background, I understand how important computer security is today. I looked for the CNS specialty program, and I could not find anything close to the program Park offers," said Michael Myers, a master's of public affairs student with a concentration in CNS. "Park University is ahead of the curve with this program. I am certain that this M.P.A. emphasis in CNS will make me a more marketable federal employee and will offer countless opportunities."

As the population ages, people need to understand the medical system. The health-care advocacy program, which is also offered online, is geared to health-care providers, nonprofit personnel working in the field, and individuals who want to learn more about health-care systems.

"Another fairly new offering is our graduate music certificate program offered through the International Center for Music. Stanislov Ioudenitch, the artistic director for the ICM, was the 2001 Gold Medalist at the Van Cliburn International Piano Competition. Park's ICM students have been consistently winning national and international competitions," said Weighill.

Park is constantly in touch with the local community. "We're located next to the Downtown Public Library, and our students appreciate the proximity and convenience that the library provides. Additionally, our International Center for Music students have performed mini-concerts in the library's beautiful space. We are also a supporter of the Kansas City Fringe Festival. This visual and performing arts festival is a prime example of the vitality that is so prevalent in downtown, where so many entities are pulling together to make things happen." ♦

THE FLEXIBILITY OF GRADUATE AND UNDERGRADUATE OPTIONS IS IMPRESSIVE.

Park University faculty and staff understand what a challenge it is to work full-time, juggle family responsibilities, and attend classes. Working adults benefit from convenient campus locations in Parkville, Independence, and downtown Kansas City.

PHOTO BY DENNIS KEIM

For more than one hundred years, the Salvation Army has been there to bring help to folks in 111 countries throughout the world and, of course, here at home. In Kansas City, the holidays are a busy time for Salvation Army volunteers, who help serve both Thanksgiving and Christmas meals with all the trimmings. In addition to the meals they serve on site, they also deliver approximately seven hundred holiday meals to homebound and elderly individuals. The Salvation Army welcomes volunteers and provides services to people of all ages, races, and cultural backgrounds.

PHOTO BY DENNIS KEIM

CHAPTER ONE: LIVING IN GREATER KANSAS CITY

On the side of the River Market Antique Mall is a magnificent, four-story mural of the Lewis and Clark Expedition of 1804. The mural commemorates a three-day stopover of the historic expedition at Kaw Point, on the Kansas side of the river, a short distance from the historic River Market area on the Missouri side.

GREATER KANSAS CITY: UNLIMITED POSSIBILITIES | 103

Swope Community Enterprises Builds Healthy Communities

"Swope Community Enterprises is dedicated to providing solutions to improve the physical, behavioral, and economic well-being of individuals, families, and communities." Twenty-one words capture the philosophy of this innovative company that provides creative answers to problems which other businesses are frequently reluctant to tackle.

"We started offering primary health services out of a church basement in 1969 as part of the War on Poverty," said E. Frank Ellis, founder, chairman, and CEO of Swope. "We were doing good work, but we soon realized improving the physical health of our neighbors wasn't enough. So, my staff and I developed a system to embrace both the physical and mental health of our patients. Again we showed promising results, but that still didn't complete the picture.
We had to address the environment too."

The answer was to create Community Builders of Kansas City, a nonprofit community development corporation that combined neighborhood homeowners with public and private entities working together to breathe new life into disenfranchised neighborhoods.

The two communities involved in the initial plan were Mount Cleveland, an area of working-poor citizens, and Sheraton Estates, a well-established community of affluent African Americans surrounded and threatened by urban decay. The work Swope Enterprises did was so successful, it was replicated by other urban communities in Kansas, Missouri, and California.

"To be a leader, you need to be an innovator," said Ellis, "to get out of your comfort zone. We do that by building from

▲ *The Shops on Blue Parkway retail center provides economic stability to the Brush Creek Corridor as well as core services and employment opportunities to surrounding communities.*

CHAPTER ONE: LIVING IN GREATER KANSAS CITY

PHOTO BY ERIC FRANCIS

the bottom up. We go to the homeowners to determine their needs, their hopes, and dreams for their neighborhoods. Getting their support has given us a groundswell of political and business support."

Swope Community Enterprises is the umbrella company for Swope Health Services, Swope Community Builders, Applied Urban Research Institute, and Swope Community Enterprise Services.

Swope Health Services provides physical and behavioral health as well as outreach services to more than forty thousand low- to moderate-income residents in a five-county area surrounding Kansas City.

Swope Community Builders offers affordable housing, neighborhood planning services, commercial development, community organizing services, and community relations programs.

Applied Urban Research Institute engages in community and economic development, urban planning, and research services for local and national clients and encourages them to reinvest in urban communities.

Swope Community Enterprise Services supplies all the routine, back-office services such as accounting, IT, human resources, and materials.

"By combining the services of all these entities," said Ellis, "we're able to accomplish our vision, which is to see 'self-empowered, healthy people in healthy communities.' Our diverse staff works in a team-based, problem-solving environment where they're encouraged to reach their full potential. We also actively involve members of the community, and we value their participation in solving problems."

Swope Community Enterprises has built a reputation for honest, trustworthy, morally responsible actions that support mutually beneficial relationships. "When we find ourselves in situations that are driven by external factors, we sometimes have to slow down, take a second look, and

(continued on page 106)

"OUR VISION IS TO SEE SELF-EMPOWERED, HEALTHY PEOPLE IN HEALTHY COMMUNITIES."

▼ *The staff in Swope Behavioral Health's Community Psychiatric Rehabilitation Center creatively engages their severely and persistently mentally ill clients.*

PHOTO BY ERIC FRANCIS

SWOPE COMMUNITY ENTERPRISES | 105

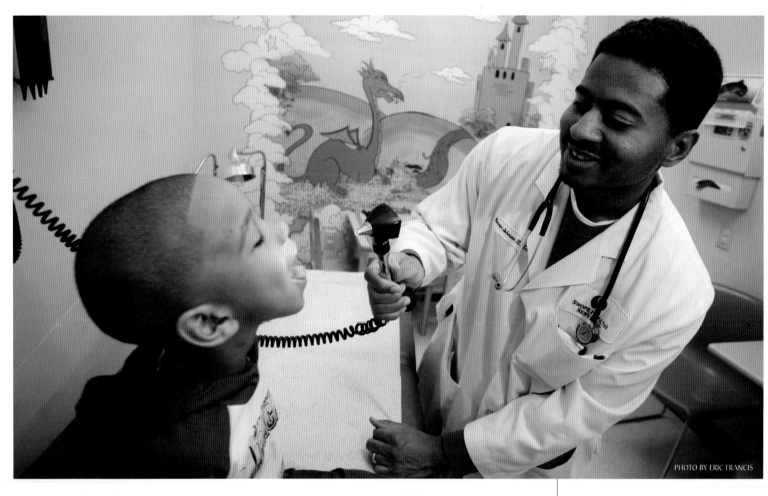

(continued from page 105)

make sure we're staying true to our core values and our mission.

"In 2004, we sold FirstGuard Health Plan, our HMO service. The resources from that sale have given us the opportunity to look for new services and products to offer. We are in the exciting process of reinventing ourselves," said Ellis.

Plans are under way to expand the core businesses by adding services for the elderly

▲ *Swope Health Services is a critical part of the Kansas City safety net of care. By providing primary medical care, including immunizations and prevention, Swope Health Services helps relieve area hospitals of the burden of providing expensive emergency care.*

◀ *In response to the critical need for dental services in Kansas City, Kansas, Swope Health Wyandotte Dental Services was established in 2006.*

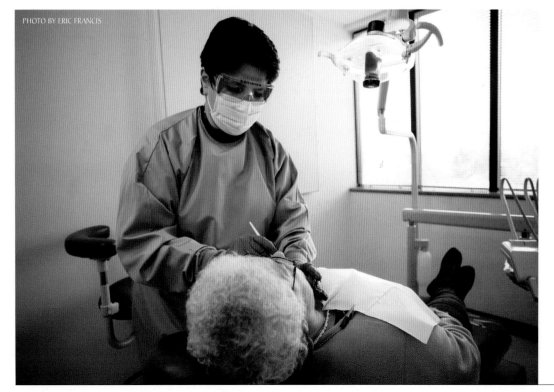

CHAPTER ONE: LIVING IN GREATER KANSAS CITY

◀ In 1999, in partnership with neighborhood residents, Swope Community Builders embarked on an $8.2 million development project to create a safe and inviting duplex community for low-income families. Mount Cleveland Heights features seventy bungalow-style duplexes with nostalgic front porches.

in both Kansas and Missouri. A workforce development effort is also in the planning stages. "This is our way to address the lack of skilled personnel in the workforce. We plan to identify high school graduates, who are not planning on going to college, and give them an opportunity to learn a trade that will lead to a successful career, not just a dead-end job. We're currently working with businesses in the area, and our target date is to roll out this plan sometime in late 2007," Ellis explained.

Another new area, which will be developed nationally, is Swope Community Enterprises Shared Services. "We will begin to offer our back-office services to other companies. As administrative duties and requirements get more and more complicated, we feel this will be a valuable service to many companies who don't want to tie up their staff doing paperwork."

The very nature of Swope Community Enterprises involves them daily in service to the community. In addition to that, employees donate countless hours to their favorite charities and sit on a number of boards. For a number of years the company has sponsored scholarships at the University of Missouri–Kansas City. Ellis himself is a trustee or sits on the board of approximately twenty local organizations.

From health care to community development and urban planning, Swope Community Enterprises is actively improving Kansas City one person, one family, one neighborhood at a time. ♦

▼ Completed in 1995, Swope Health Central serves as both a comprehensive health resource and an economic engine for community revitalization in south-central Kansas City.

PHOTO BY MARIO MORGADO

E. Frank Ellis, chairman of Swope Community Enterprises, speaking at the Greater Kansas City Chamber of Commerce Annual Dinner. The Chamber is just one of the many organizations in Kansas City that benefits from the volunteer hours donated by Swope associates.

PHOTO BY DENNIS KEIM

Kansas City's hometown railroad, Kansas City Southern, constructed a new headquarters in Kansas City in April 2002, remaining close to its downtown roots established in 1887.

PHOTO BY THOMAS S. ENGLAND

PHOTO BY THOMAS S. ENGLAND

CHAPTER ONE: LIVING IN GREATER KANSAS CITY

Built in Brownsville, Pennsylvania, in 1853, the ill-fated *Arabia* was destined to survive only three years of service on the waterways. Equipped with twenty-eight-foot-tall paddlewheels, the *Arabia* could navigate a swift river current, fully loaded, at a speed of six miles per hour or more, performance that fortified her reputation as a speedy, comfortable, safe ship. But on September 5, 1856, after a short stopover in Westport Landing, site of today's Kansas City, Missouri, the boat quickly sank after its hull was pierced by a solidly rooted tree snag. Bob Hawley, along with family (shown here with wife Florence and son David, above) and friends, raised the *Arabia* in 1988. From porcelain to pistols, trousers to tools, the items found in the muck of the ruins can be found at the Arabia Museum. In addition to the six-ton stern section, carefully preserved for display, a full-sized reproduction of the *Arabia*'s main deck is on display for visitors to see.

GREATER KANSAS CITY: UNLIMITED POSSIBILITIES | 109

◀ *Dr. Sarah Schwartz, senior chemist, and Dr. Dean Gray, section manager and expert in botanical studies, conduct plant biomarker research that has far-reaching applications in plant science and health studies.*

Midwest Research Institute: Benefiting Society through Scientific Research

When M&Ms melt in your mouth and not in your hand, thank Midwest Research Institute (MRI), the scientific research organization that, in the 1950s, developed the process to apply the M&M candy coating to the chocolate.

By the 1960s, MRI, headquartered in Kansas City, had expanded from regional industrial projects to government and medical contracts: entering the space race through NASA research projects, developing anticancer compounds for the National Cancer Institute, and assisting the Department of Defense with the safe destruction of the nation's stockpile of chemical weapons. Founded in 1944 by nine Kansas City business leaders, including J. C. Nichols, this diverse institute, with a mission to benefit society, is now internationally acclaimed for its scientific and technological advances.

"MRI's research focus has undergone a significant evolution," said Linda Cook, director of communications. Following September 11, 2001, MRI escalated its work on national defense with designs such as the SpinCon air sampler that detects trace levels of biological agents. Other activities include an emphasis on proteomics to study various diseases, test pharmaceuticals, and conduct research in the areas of energy, agriculture, and animal health.

MRI maintains close ties to its founding city through the sponsorship of such organizations as Science Pioneers to encourage student involvement in science and technology. A Science Pioneers' mainstay—the Greater Kansas City Science and Engineering Fair—recently celebrated its fiftieth year. Essentially, MRI has evolved to meet the needs of an ever-changing world, and its dedicated staff plans to continue doing so through future innovations in energy, defense, and life sciences. ♦

MRI HAS EVOLVED TO MEET THE NEEDS OF AN EVER-CHANGING WORLD.

CHAPTER ONE: LIVING IN GREATER KANSAS CITY

Boasting restaurants, businesses, and services unique and consistent with the original neighborhood's charm and architecture, the Brookside shopping district is the first outdoor shopping center built in south-central Kansas City, Missouri, and the focal point of the surrounding historic neighborhood. Brookside's unique flavor, a mix between nostalgic and trendy, earns it the reputation of the place to live or at least visit. J. C. Nichols, a developer ahead of his time, designed the area in the early 1900s with the concept of integrating a commercial district to support the surrounding housing. "It's a city within a city, with churches, schools, and shops," explained Marti Lee, executive director of the Brookside Business Association. "Today, Brookside is used as a model for new developments throughout the country." With more than seventy businesses, beautiful homes, and residents who still enjoy sitting on their front porches, Brookside welcomes visitors to its local parades and events, including a nationally acclaimed annual art fair.

The 129,000-square-foot Johnson County Sunset Drive Office Building, Olathe, opened in 2006 with more than three hundred county employees from seven departments and agencies working in the facility. It is the first building in Johnson County certified LEED (Leadership in Energy and Environmental Design) Gold by the U.S. Green Building Council. The two-story Sunset Office Building is only the second LEED Gold facility in the Kansas City metropolitan region and the state of Kansas.

CHAPTER ONE: LIVING IN GREATER KANSAS CITY

PHOTO BY THOMAS S. ENGLAND

The Central Library in downtown Kansas City is the major resource library for the system, which includes nine branch libraries. In addition to being a source for serious research, the library has a nonstop flow of programs. Visitors may be surprised to hear live music—furnished here by local musicians Denny Osburn and Mark Montgomery—floating down the halls, or recognize well-known authors talking about their work. The library also offers Mouse Basics for computer enthusiasts like young Jaslyn, who seems to be totally engrossed in her work.

ALL PHOTOS BY DENNIS KEIM

Kansas City definitely lives up to its fame as the City of Fountains. The first fountain, erected in 1904, was designed with pools at different levels to provide water for people, horses, dogs, and birds. Today Kansas City has approximately two hundred fountains—more than any city except Rome. From the landmark J. C. Nichols Memorial Fountain with its four equestrian figures representing four famous rivers of the world, to the playful Children's Fountain with the six bronze sculptures of children playing in the water, each fountain tells its own story and has its own secret. (Tom Corbin, who designed the Children's Fountain, used local children as models for his sculptures.) There are fountains celebrating heroes such as veterans and firefighters, as well as numerous small fountains tucked away in private backyards. The best may be yet to come as the City of Fountains Foundation plans for the future.

a. Brush Creek Fountain, runs through Country Club Plaza
b. Barney Allis Plaza Fountain, Twelfth and Wyandotte streets
c. Children's Fountain, North Kansas City, North Oak Trafficway and Missouri 9
d. J. C. Nichols Plaza Fountain, Country Club Plaza, Forty-seventh Street and Nichols Parkway
e. Muse of the Missouri, Eighth and Main streets
f. William T. and Charlotte Crosby Kemper Memorial Fountain, Tenth and Main streets
g. Mermaid Fountain, Broadway and J. C. Nichols Parkway
h. Pomona Fountain, 320 Ward Parkway and Wornall Road, on Country Club Plaza

GREATER KANSAS CITY: UNLIMITED POSSIBILITIES | 115

CHAPTER ONE: LIVING IN GREATER KANSAS CITY

Thousands of people annually come from miles around to enjoy the festivities during the Mayor's Christmas Tree Lighting in Crown Center Square. Towering as much as one hundred feet, the centerpiece Christmas tree is hand-picked each year and brought to the city for its holiday season decoration. The ceremonial flipping of the switch by the mayor and a celebrity guest illuminates the thousands of lights adorning the tree and the center as a kickoff to the holiday season. The celebration has included spectacular events like carolers, fireworks, and a fantastic water show spouting from the Crown Center Square Fountain's nearly one hundred water jets and shooters. Since the first Mayor's Christmas tree was erected in 1908, it has served as a reminder to give to the Mayor's Christmas Tree Fund, which benefits the area's less fortunate. Each year since 1987, when the festivities have concluded, wood from the tree is then made into the Mayor's Christmas Tree Ornament, designed by Hallmark, and sold as part of the mayor's holiday fund-raising efforts.

Working in KC
Greater Kansas City

A lot has happened around Kaw Point since the Lewis and Clark Expedition landed there on June 26, 1804. The expedition spent three days at the place where the Kansas River flows into the Missouri River, resting up and preparing for the journey westward. To commemorate the historic stayover, a park was dedicated at Kaw Point during a bicentennial celebration in 2004. Today, Kaw Point offers some of the area's most picturesque views of the Kansas City skyline, a pioneering town at the crossroads of the nation that became a modern city spanning two rivers and two states.

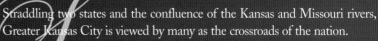

Straddling two states and the confluence of the Kansas and Missouri rivers, Greater Kansas City is viewed by many as the crossroads of the nation.

Since the first trading post established here in 1821, Kansas City has been a place that welcomes entrepreneurial thinking. While the waterway was the main form of cargo transport in those early years, a steady flow of settlers brought commerce along the trails as they made their way westward. Transcontinental rail also brought travelers and cattle to a city that is still the nation's second-largest rail center, home to Kansas City Southern and twelve rail lines moving passengers and freight.

Kansas City's central location also makes it easy to connect to anywhere by air via both international and municipal airports, nearly a dozen air freight carriers, small charters, and corporate aircraft operations. Fourteen interstates or U.S. highways run through the city or connect to larger roadways, making it one of the nation's top trucking and intermodal freight hubs, served by more than two hundred freight carriers.

With nearly 2 million residents, the Kansas City metropolitan area is home to ten Fortune 1000 companies. But it is also considered to be one of the nation's best places for small business start-ups, with more than fifty thousand companies employing fewer than one hundred personnel. Some of the world's greatest success stories have started in Kansas City, including Hallmark Cards Inc., American Century Investments, and Burns & McDonnell.

Kansas City is also the nation's leader in the animal health and nutrition industry, accounting for nearly 30 percent of the world's $14.2 billion market. Other industries driving the area's economy range from retail and finance, to health care and telecommunications, to manufacturing and construction.

Keeping future workforces educated is the task of more than twenty universities, colleges, and professional schools, including Kansas City University of Medicine and Biosciences, specifically focused on keeping tomorrow's researchers and scientists at the cutting edge.

Attractive in its architecture, Kansas City is also home to companies that create spaces large and small. From nationally known JE Dunn Construction to the global reach of Black & Veatch, the area's engineering and design firms leave a Kansas City footprint across the nation and abroad.

A central location, diverse economy, progressive business outlook, educated workforce, and so much to see and do when the working day is over . . . these are the things that make Greater Kansas City a great place to do business.

AMERICAN CENTURY INVESTMENTS	132
ARMSTRONG TEASDALE LLP	186
BANK MIDWEST	154
BKD, LLP	208
BLACK & VEATCH CORPORATION	158
BLACKWELL SANDERS PEPER MARTIN, LLP	146
BURNS & MCDONNELL	122
CITY OF BLUE SPRINGS	190
FOGEL-ANDERSON CONSTRUCTION CO.	164
GOULD EVANS	170
GREATER KANSAS CITY CHAMBER OF COMMERCE	128
HDR ENGINEERING, INC.	140
JE DUNN CONSTRUCTION	180
KEY COMPANIES & ASSOCIATES LLC	202
LATHROP & GAGE L. C.	142
LEWIS, RICE & FINGERSH L. C.	176
POLSINELLI SHALTON FLANIGAN SUELTHAUS PC.	150
SHOOK, HARDY & BACON LLP	198
TOP INNOVATIONS, INC.	194

GREATER KANSAS CITY: UNLIMITED POSSIBILITIES | 119

When the artist known as Stretch (a.k.a. Jeff Rumaner) starts to create one of his signature sculptures, the sparks literally fly. He works in both glass and metal, and his works range in size from small pieces to those that dominate the landscape. The larger pieces have been described as "a dialogue between the viewers and the architecture." Stretch is a graduate of the Kansas City Art Institute and is a vital part of the Kansas City art scene. He has also made numerous television appearances on shows such as ABC's *Extreme Makeover, Home Edition*. In addition to his artwork, Stretch became a restaurateur when he opened Grinders in 2004. One of his latest ventures was creating a large-scale sculpture for the new H&R Block Center downtown.

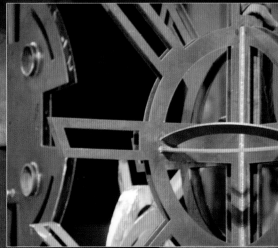

CHAPTER TWO: WORKING IN GREATER KANSAS CITY

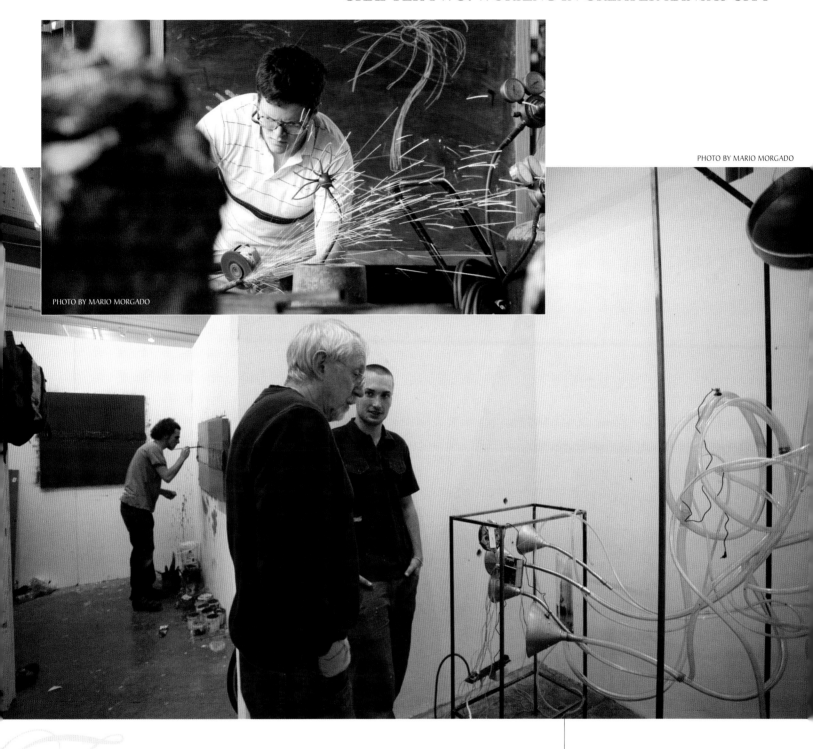

Founded in 1885, the Kansas City Art Institute (KCAI) in Kansas City, Missouri, offers bachelor's degrees in fine arts such as painting, sculpture, and photography or design (animation and graphics). Known for its rigorous standards, schools around the country use KCAI as a teaching model. In 2005, the college began a multiyear, multimillion-dollar plan to enhance the campus, including its newest facility, the Lawrence and Kristina Dodge Painting Building. In that same building, Nathan P. Davies, a student painter who has just received a professor's critique, stands proudly with his brushes and his project, a triptych (a set of three painted panels). Surrounded by the works of other students in a Sculpture Building classroom, Tex Jernigan, a senior in the sculpture department, works on a maquette (a small three-dimensional version of a project), which is also sketched on the blackboard behind him. Travis Smorstad, a senior, receives an appraisal of his artwork from Warren Rosser, the William T. Kemper Distinguished Professor and chair of the painting department at KCAI.

Burns & McDonnell Represents KC to the World

An airport in Saudi Arabia. A power plant in China. Wastewater facilities in Mexico. The common denominator to all of these facilities? Burns & McDonnell designed them.

It all started in 1898 with two extraordinary engineers: Clinton S. Burns and Robert E. McDonnell. Using the methodical precision of their profession, they conducted a study and determined that of the largest U.S. cities, Kansas City had the greatest number of municipalities within two hundred miles lacking basic infrastructure—the focus of their planned engineering practice.

Despite never having set foot in the area, the two men came to Kansas City in April 1898 and began bringing sanitation, clean water, and reliable power to the rural communities of the region.

Today Burns & McDonnell has more than twenty-three hundred employees in offices nationwide providing engineering, architecture, construction, environmental, and consulting services in every state and thirty-three countries. They have completed projects on every continent except Antarctica.

"At Kansas City Power & Light, we've partnered with Burns & McDonnell on

▲ *Burns & McDonnell designed this concrete hangar in Guam. The hangar was built for forward deployment of the B-2 (stealth) bomber. The hangar was constructed with concrete so it could withstand the typhoons typical for this area.*

CHAPTER TWO: WORKING IN GREATER KANSAS CITY

ident and CEO. "It defines the culture here at Burns & McDonnell. We don't just call ourselves employee-owned; we *are* employee-owned. It may sound like a cliché, but it's true. Our employees do go the extra mile for our clients.

"A key part of our long-term business model is diversification. We started in 1898 with three practice areas: water treatment, power generation, and sanitation. Since then, we have broadened our range, and today we offer more than 350 distinct services. This strategy allows us to weather inevitable market fluctuations."

The range of projects and locations in which Burns & McDonnell is involved is staggering. One only has to visit downtown Kansas City to see the extent to which the company is engaged in the area's rebirth. From the Sprint Arena to the *Kansas City Star* press pavilion, from the new entertainment district to basic infrastructure, the Burns & McDonnell "touch" is evident throughout downtown and the entire metro area.

(continued on page 124)

projects critical to both our business and to the community in general," said William Downey, president of KCP&L. "I've always been impressed that, even though it handles projects around the world, Burns & McDonnell plays a significant role as a prominent corporate citizen right here in Kansas City."

One of the primary factors that distinguishes Burns & McDonnell is its 100 percent employee ownership. Each employee receives stock on an annual basis, and all reap the benefits of ownership. There are no outside stockholders.

"The extra dedication of employee-owners makes all the difference when we serve our clients," said Greg Graves, pres-

"IT MAY SOUND LIKE A CLICHÉ, BUT IT'S TRUE. OUR EMPLOYEES DO GO THE EXTRA MILE FOR OUR CLIENTS."

▼ *Burns & McDonnell was the original designer and builder of Kansas City International Airport. The firm's most recent major assignment was as program manager for the Terminal Improvement Project at the airport.*

◀ *Burns & McDonnell designed the $134 million Lake Fort Smith Dam and Reservoir Enlargement project in Arkansas. It is the largest earthen dam recently constructed anywhere in the United States. Reservoir capacity has been tripled, enabling the city of Fort Smith and neighboring communities to easily meet projected growth in water demand until at least the year 2050.*

(continued from page 123)

Nationally, Burns & McDonnell is making its mark and exporting Kansas City know-how. A few examples:

• Burns & McDonnell has been a key player in restoring the Florida Everglades to its former glory. Using an innovative process, the firm has designed a system to dramatically clean up much of the water entering the Everglades.

• The firm is planning, designing, and building many defense-related facilities, including the high-tech support facilities for the Joint Strike Fighter, a projected $200 billion project that may be the last manned-fighter aircraft built.

• Burns & McDonnell is at the forefront of the effort to rebuild the electrical grid after the massive 2003 power blackout in the Midwest and Northeast. Most notably, a $1.3 billion underground and overhead transmission project in Southwest Connecticut is eliminating a major bottleneck in that area.

"The entrepreneurial spirit at Burns & McDonnell is taking us to new heights every day," Graves said.

Health-care facilities are yet another field of interest for Burns & McDonnell. "This is a complex, ever-expanding field, and it requires a highly trained staff like ours," said Graves.

Although Burns & McDonnell has a global presence and is involved in a multitude of high-profile projects, employee-owners still find time for hands-on involvement in Kansas City.

"Our employees' record of volunteerism is excellent. All our officers serve on at least one civic board, and a great number of employees participate in events such as Christmas in October and various fund-raising walks. We are also active in

CHAPTER TWO: WORKING IN GREATER KANSAS CITY

◄ Burns & McDonnell has touched many of the projects associated with the recent renaissance of downtown. For the Sprint Arena, Burns & McDonnell acted as program manager on behalf of the City of Kansas City.

▼ Iatan 2 (bottom left), an 850-megawatt coal-fired facility located just northwest of Kansas City, is being designed by Burns & McDonnell. Kansas City Power & Light will be majority owner and operator of the facility in partnership with Aquila, Inc., Empire District Electric, Kansas Electric Power Cooperative (KEPCO), and Missouri Joint Municipal Electric Utility Commission (MJMEUC). After Iatan 2 goes online and environmental upgrades are completed at the adjacent Iatan 1 facility, power capacity at the Iatan campus will more than double. There will also be substantial improvement in air emissions. Another way Burns & McDonnell is making the air you breathe cleaner is with its clean fuels projects at refineries. The firm has been providing engineering and construction services to reduce sulfur in your fuel, such as this project (below) for ConocoPhillips.

many youth education efforts, such as MathCounts, which encourages learning math and science in middle schools," Graves explained.

Whether around the world, across the country, or in downtown Kansas City, Burns & McDonnell touches many people's lives. From the water they drink, the air they breathe, and the airports and roads they travel, to the electricity that powers their homes and businesses, Burns & McDonnell is there making life better for all. ♦

PHOTO BY MARIO MORGADO

The Commerce clock has marked time since 1953 when it was installed on the Commerce Trust Building at Tenth and Walnut, in Kansas City, Missouri. Frances Reid Long founded the Commerce Bank in 1865, shortly after he arrived in the growing city with ten thousand dollars and a desire to invest it. In 1906, Commerce Bank, then known as the National Bank of Commerce, planned a building to accommodate the bank and the newly chartered Commerce Trust Company. The building, designed by Jarvis Hunt and constructed in 1907–08, is listed on the National Register of Historic Places. The five-feet-by-eight-feet cast-bronze and copper Commerce clock weighs 3,330 pounds and stands eleven feet from ground level. Designed by the Livers Bronze Company in collaboration with IBM, the clock's original cost was approximately sixteen thousand dollars. In 2002, the building underwent a major renovation, including restoration and preservation of aspects of the exterior and main lobby.

CHAPTER TWO: WORKING IN GREATER KANSAS CITY

Expansion of Bartle Hall at the Kansas City Convention Center is an architectural marvel demonstrating some of the structural feats that dot the city's landscape. Because construction of the addition was limited by space issues among existing downtown structures, it was designed to span sixteen lanes of traffic over the adjacent highway. Building the space required pouring concrete columns in place on the median and shoulders of the highway and supporting the roof with a system of cables.

◀ *Centurions participate in community service work throughout their tenure in the two-year program. During the holidays, members of the Centurions program volunteer to ring bells for the Salvation Army. Pictured (left to right) are Alex Wendel of Fleishman Hillard, Laura LaPorte of the Kansas City Public Library, and Dan Friedrich of Grant Thornton.*

PHOTO BY BRUCE MATHEWS

The Chamber: It Works. In Kansas and Missouri.

In a city straddling two very different states, the Greater Kansas City Chamber of Commerce is on a mission to facilitate regional cooperation.

With members in both Missouri and Kansas, The Chamber works as the voice of business to implement positive change for the greater good of the region as a whole.

For instance, by leading the campaign to resuscitate Kansas City's historic train depot, Union Station, The Chamber was instrumental in saving a structure now commonly used as a gathering place by the community.

Because education is a priority issue for The Chamber, the organization works with legislators, educators, and business leaders in both states to navigate issues that jeopardize the stability of school systems and to connect entities that shape the needs of future workforces.

Along with the Kansas City Area Development Council and the Kansas City

Area Life Sciences Institute, The Chamber takes the lead in legislative and public policy issues associated with the Kansas City Animal Health Corridor, an idea launched by one of its former chairmen. To date, companies in this corridor, extending from Manhattan, Kansas, to Columbia, Missouri, account for almost one-third of sales in the multibillion-dollar global animal sciences market.

The Chamber also lends its voice to public transit and air quality issues—for example, working to maintain the region's compliance with EPA air quality standards.

As a progressive organization, The Chamber recognizes the value of best practices outside its own jurisdiction and capitalizes on those through its Leadership Exchange, a program through which local business, elected, and civic leaders visit other cities to study those communities' best practices. An offshoot of the Leadership Exchange, the Regional Alliance studies issues and success stories within the Greater Kansas City area, spurring development of initiatives like the ArtsKC Fund, a regional, workplace-based giving program to benefit the arts.

While involvement in public policy is a key function for The Chamber, the organization also connects people through programs and services that came about as a result of member input. Asked what they needed from their Chamber, members overwhelmingly replied, "Help us get smarter, better connected, more visible."

Therefore, in answer to members who want to gain knowledge and grow their bottom lines, The Chamber coordinates low-cost seminars, business and leadership development programs, and specially discounted benefits. To help members connect with other businesses, elected officials, and customers, The Chamber coordinates relationship-building opportunities such as Business Before and After Hours gatherings, active committees, workforce development programs, and more. And through its publications, awards, and public engagements, The Chamber provides members with an array of prospects for more visibility. Collectively, The Chamber's activities reveal it to be an organization driven to make Greater Kansas City and the region an increasingly greater place to call home. ♦

THE CHAMBER WORKS AS THE VOICE OF BUSINESS TO IMPLEMENT POSITIVE CHANGE FOR THE GREATER GOOD OF THE REGION AS A WHOLE.

◄ The Kansas Citian of the Year Award is The Chamber's highest honor, and it is bestowed at Annual Dinner, a gala event held each year on the Tuesday evening before Thanksgiving. Past Kansas Citians gather each year to congratulate the new honoree. Back row (left to right): Shirley Helzberg (2001), Robert Kipp (2000), the Rev. Emanuel Cleaver (1999), and Anita Gorman (1992); front row (left to right): Barnett Helzberg (2001), Jack Steadman (1988), Drue Jennings (2006), and Carol Marinovich (2005).

PHOTO BY MARIO MORGADO

In most cities, residents never get the thrill of seeing inside a pressroom and watching the gigantic, high-speed presses as they roll. However, that's not the case with the *Kansas City Star*. Its massive $199 million, two-block-long, glass-enclosed printing and distribution plant contains four sixty-foot presses on view daily. The building, which took four years to build, was part of a major downtown revitalization. The *Star* has been part of the community since it was founded in 1880 as the *Kansas City Evening Star*.

Kansas City, Missouri, has emerged over the past two centuries from being wilderness territory of the Louisiana Purchase to an eclectic city famous for a skyline that includes such structures as the Bartle Hall Convention Center, the Power and Light Building, and the Liberty Memorial. On the thirteenth of Main Street's seventy blocks sits the stunning, oblong-shaped, seventeen-story H&R Block World Headquarters, completed in 2006. H&R Block is a full-service tax preparation and financial management firm started by two Kansas City brothers in 1946 that today serves more than 20 million clients annually. The new green-glass-enclosed headquarters displays the artwork of local artisans, from small paintings to huge wall hangings in its tower, and most of the lobby and atrium serves as a downtown art gallery. Because of its unusual shape, the work spaces and furniture for H&R Block's sixteen hundred employees were custom-designed.

PHOTO BY MARIO MORGADO

Originally constructed in 1952–54, the Paseo Bridge handles ninety-five thousand vehicles per day, making it the busiest bridge in Kansas City, Missouri. It carries a total of four lanes over the bridge: Interstates 29 and 35, and U.S. Route 71. Serving as the main connection between Kansas City and the Northland (the area of Kansas City north of the Missouri River) has taken a toll on this historic suspension bridge. Rehabilitated in 1984, the bridge was shut down for repairs in 2005, with alterations designed to prolong the life of the structure for the next ten years. The length of the Paseo totals 1,831.6 feet, with its main span of 616 feet traversing the "Wide Missouri," or the "Big Muddy," as the river is often called.

As the primary design/build electrical contractor for Kansas City's new IRS Service Center, Mark One Electric, worked closely with general contractor JE Dunn Construction to design and implement all power and lighting for the state-of-the-art processing facility. A major consideration throughout the project was working to meet the requirements of the Leadership in Energy and Environmental Design (LEED) program, whose principal objective is to reduce light pollution and lower energy costs. One of the building's outstanding features as a result is its efficient and light-filled main atrium. Created by eliminating two stories of the old post office space, the atrium is lit by skylights during the day, energy efficient lamppost style lighting by night.

American Century Investments Succeeds by Making Others Successful

Since its founding by James E. Stowers Jr., American Century Investments has been guided by the belief that the ultimate measure of its success is the contribution the firm makes to the success of its investors and clients. In pursuit of this goal, the firm utilizes the best research, tools, and technology available; executes disciplined and repeatable investment processes; and strives to build an organization with strong values and principles.

Stowers achieved his mission to improve people's financial position with one simple investment philosophy: "money follows earnings." He believed that over time investors in equity markets naturally gravitate to companies with high revenues, and he focused his portfolios on those companies exhibiting accelerated growth.

By the 1970s, Stowers recognized the potential for computers to streamline his process. Frustrated by the inability of computer experts to write an appropriate program, he took a computer course and taught himself COBOL. While at a conference in Quebec City, he woke up with an idea. Tiptoeing into the hotel bathroom so not to wake his wife, he spent the next several hours writing the program himself.

This kind of innovative thinking has defined American Century® since 1958. That was when Stowers, a young insurance

▲ *American Century's Investor Centers enable investors to meet face-to-face with an experienced representative. Investor services range from individual and brokerage account transactions, to establishing long-term investment plans for goals such as retirement and college savings. Investors also conduct business with American Century by phone and online at www.americancentury.com.*

broker, realized the wealth-building potential of mutual fund investing. Starting with only one hundred thousand dollars and two funds, he grew American Century into one of the most trusted investment companies in the nation.

Today, American Century's teams of investment professionals manage assets for individual, corporate, and institutional clients. Investment disciplines include fixed-income, value, quantitative equity/asset allocation, growth, and international equity. Each is supported by extensive in-house research and analysis, as well as time-tested investment approaches. American Century offers its investment management expertise in mutual funds, sub-advisory accounts, institutional separate accounts, and co-mingled trusts.

"Because our company's long-term success depends on making our customers successful, we intend to continue building American Century's capacity to meet the investment management needs of investors and clients through a broad offering of high-quality, actively managed investment strategies," says president and CEO Jonathan Thomas. "Some investors do business with us directly, and some investors obtain our investment management through a third party such as a bank, broker, advisor, or insurance company. We also serve institutional endowments and retirement plan sponsors. Once we have established a client relationship, we are totally committed to making our client successful."

Bolstered by the belief that the best way to grow its clients' assets is by investing for the long term, American Century has long stood against the practice of short-term trading in mutual funds. The company vigorously works to stop abusive trading practices by devoting staff to investigating signs of abuse, implementing redemption fees for early sales, utilizing fair-value pricing, and severing ties with offenders.

"As a premier investment manager, our products and services distinguish themselves positively in the marketplace every day," says Thomas. "What we do is very measurable and easy to compare to our competitors."

Measurable results—repeatable, consistent, and long-term—are what lead stock and mutual fund ratings companies like *Morningstar* to rate American Century among the nation's best firms in the industry. American Century is consistently recognized as a top-notch employer. The company's eighteen hundred employees, the majority in Kansas City, give their employer such high marks that *Fortune* magazine continues to rate American Century among its 100 Best Companies to Work For.

Operating with the utmost integrity forms the basis for the company's culture.

(continued on page 134)

FORTUNE MAGAZINE CONTINUES TO RATE AMERICAN CENTURY AMONG ITS 100 BEST COMPANIES TO WORK FOR.

▼ *American Century is committed to improving financial literacy for students of all ages. Employees volunteer to teach financial fundamentals, and educators can access free financial education resources online. Tips for Kids® and Tips for Life® are signature programs, available at www.tipsforkids.com.*

PHOTO BY DAVE GILLISPIE

AMERICAN CENTURY INVESTMENTS

◀ *American Century Investments' founder James E. Stowers Jr. and his wife Virginia established the Stowers Institute for Medical Research in 1994. The world-class medical research facility employs more than four hundred scientists from around the world and is devoted to discovering cures for gene-based diseases.*

(continued from page 133)

"We believe American Century is for investors and clients who demand results and have strong values," Thomas continues. "We believe in doing things right *and* doing the right things."

This philosophy guides American Century to contribute its time and assets to a number of worthy community enrichment programs via the American Century Foundation, corporate charitable sponsorships, and company-wide volunteerism. One such initiative has been its financial literacy program. These curricula leverage the skills that American Century has as a financial company and offer those skills to help middle and high school students. The online, hands-on program helps students learn about managing their personal finances as well as school-to-career issues. Both programs are free for any educator to use. American Century developed them as a way to give back to the community.

Motivated by their own experience with cancer and a desire to give something back to the community that nurtured American Century into a successful enterprise, Stowers and his wife Virginia

◀ *The investment firm's Kansas City headquarters is located near the beautiful Country Club Plaza.*

CHAPTER TWO: WORKING IN GREATER KANSAS CITY

founded the Stowers Institute for Medical Research in 1994. Located on a ten-acre facility within view of American Century's Kansas City headquarters, the institute is home to dozens of top scientists studying how genes can be altered to prevent or slow disease. The institute is supported by a $2 billion endowment, primarily made up of securities and cash gifts from the Stowers family. ♦

▲ *The American Century Investments trading room is a hub of activity where traders keep a constant watch on the stock market.*

◀ *American Century Investments proudly sponsors the Plaza Art Fair, which is recognized as one of the top-five fine art fairs in the country. The three-day event attracts 230 artists, thirty-eight arts organizations, and three hundred thousand visitors to the Country Club Plaza each September. More than seventy-five American Century employees volunteer their time each year to help with the fair.*

AMERICAN CENTURY INVESTMENTS | 135

CHAPTER TWO: WORKING IN GREATER KANSAS CITY

At the Mahaffie Stagecoach Stop, visitors can experience life as it was over a hundred years ago. As the last remaining stagecoach stop on the historic Santa Fe Trail still open to the public, Mahaffie features three original remaining buildings: the Mahaffie Family Home/Stagecoach Stop, two-story Ice House, and Wood Peg Barn. Located in Olathe, just outside of Kansas City, Mahaffie is preserved as a living history site by a team of enthusiastic interpretive specialists, like Laura Daughtery, who in the summer months demonstrates for school groups the basics of life on a farm. Geoffrey Bahr not only cares for the town's livestock, he also works as its blacksmith and leather craftsman. Wayne Maltbie, Mahaffie's seasonal stagecoach driver, provides children like Lauren Fuller and John Bulter with a firsthand experience in old-fashioned horsepower.

GREATER KANSAS CITY: UNLIMITED POSSIBILITIES | 137

138 | GREATER KANSAS CITY: UNLIMITED POSSIBILITIES

CHAPTER TWO: WORKING IN GREATER KANSAS CITY

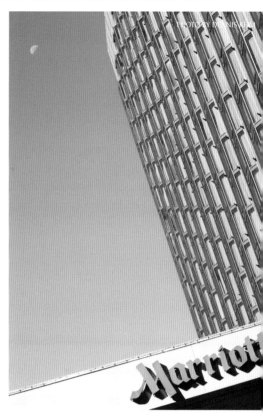

Even a quick tour of Kansas City's architecture reveals the city's history as a boomtown both past and present. Over a century ago, in 1888, the first skyscraper rose over the Kansas City skyline, leading to the construction of some fifty structures towering ten stories or more over the city. Between 1920 and 1940, the cityscape became dotted with art deco-style structures, including the distinctive Fidelity National Bank and Trust Company (far left). Missouri's tallest apartment building at thirty-five stories high, its original address was 911 Walnut Street. After the events of 9/11, its address was changed to 909 Walnut Street. Notable modern-day buildings include One Pershing Square and Two Pershing Square, H&R Block World Headquarters, the Kansas City Star building, and One Kansas City Place, Missouri's tallest habitable structure. Today, the city's downtown area is undergoing a renaissance, with the revamping of many of its beautiful and remarkable structures to become mixed-use residential and retail buildings. Such is the case with the Stuart Hall Stationery Company. Built in 1910 as a manufacturing plant for what is known today as Nabisco, it now houses 115 loft apartments.

PHOTO BY SCOTT DOBRY PICTURES

HDR: One Company, Many Solutions

HDR is a national architectural, engineering, and consulting firm specializing in serving its local areas. "We want everyone to know that we are dedicated to being an enduring part of the Kansas City community," said Donald Curtis, department manager. "We have a lot of talented minds working to solve whatever problems face our community now and into the future.

"Whether the project is transportation, drinking water and wastewater management, or the environment, we always strive for 'leap-ahead technology.' We help our clients address challenges that might not have existed a decade ago."

A good example is the I-635/I-70 interchange. "I-635 carries sixty thousand vehicles a day, and I-70 carries ninety thousand. Our job was not just to help the DOT move more traffic, but to move it more safely," said Curtis.

Another example is the BNSF Railroad Gardner Intermodal Facility. "As the hub for domestic and international container shipments arriving by train for transfer to trucks for regional distribution, this facility will have a major economic impact on our area."

Since HDR moved to Kansas City, the firm has grown rapidly, with 80 percent of its business coming from repeat clients. As HDR has grown, so has its civic involvement. HDR employees are improving the quality of life both at work and in the community. The company continually exceeds its United Way donation goals, while also supporting the Missouri Adopt-a-Highway program and SAFEHOME, a shelter dedicated to eliminating domestic violence.

Clearly HDR is committed to making Kansas City home. ♦

▲ *(Left to right) Alex Wright, of HDR Engineering, Inc., and Judy Penner, from Kansas City Missouri Parks and Recreation, get a couple of helping hands from Bill Horner and Amanda Schulte from HDR Engineering, Inc., who volunteered to assist in planting flowers at Loose Park in Kansas City.*

"WE HELP OUR CLIENTS ADDRESS CHALLENGES THAT MIGHT NOT HAVE EXISTED A DECADE AGO."

PHOTO BY MARIO MORGADO

CHAPTER TWO: WORKING IN GREATER KANSAS CITY

The result of a massive community-based fund-raising effort that brought in over $2.5 million in 1919 alone, the Liberty Memorial complex was intended as a timeless, living expression of gratitude for all those who gave their lives in service during World War I. Completed on November 11, 1926, the complex includes two museum buildings, a dedication memorial with bronze busts of the war's five Allied leaders, and a series of courtyards and stairways linking it all together. Two of the complex's most impressive structures are a 217-foot-tall memorial tower with carved stone sculptures and the *Great Frieze*, a commentary on the progress from war to peace. On August 4, 1998, Kansas City voters passed overwhelmingly a one-half-cent sales tax for eighteen months, which raised $30 million for the restoration of the memorial and $14.7 million to endow its future maintenance. Restoration began in 2000 and was completed for the rededication and reopening of the Liberty Memorial in May 2002. Also on-site at the memorial is the nation's official World War I Museum, a state-of-the-art venue that brings the history of the conflict to life via interactive exhibits, original artifacts and documents, and lectures and discussions.

PHOTO BY ALAN S. WEINE

Lathrop & Gage:
A National Law Firm Committed Locally

With more than 130 years of experience, Lathrop & Gage has expertise in virtually every area of the law. The firm has nearly doubled in size over the past decade, but continues to represent its very first client, BNSF Railway Co., in a relationship that began in 1873—a true testament to the firm's adaptability and commitment to client service.

With more than 270 attorneys specializing in twenty-five practice areas, Lathrop & Gage combines the depth and strength of a national firm with the hands-on service, management style, and aggressive competition of a smaller firm. In fact, the firm was recently named one of the legal industry's Best-Kept Secrets by *Of Counsel* magazine for its high-quality work at midwestern rates.

Lathrop & Gage's national practice advises clients on every major area of legal representation:

- **Litigation:** The firm has tried cases nationwide (see www.beentherewonthat.com) and has special expertise in class-action and complex litigation.
- **Corporate:** Lathrop & Gage's business division has been recognized by *Corporate Board Member* as among the top corporate practices in the country.
- **Intellectual Property:** The firm's IP division is one of the largest and most

▲ *Working for Jackson County, Lathrop & Gage's Jack Craft, Steve Mitchell, and Tom Stewart (pictured left to right) negotiated and drafted new stadium leases for the Chiefs and Royals in 2006.*

experienced groups in the Midwest, with more than forty lawyers who are go-to talent in patent, trademark, copyright, IP litigation, trade secrets, and licensing.

The firm is proud to count many of the city's leading companies among its clients, including AMC Entertainment and YRC Worldwide. Headquartered in Kansas City, the firm has offices in Overland Park and seven other cities nationwide.

COMMITTED TO KANSAS CITY

One of Kansas City's first mayors, Turner Gill, was an attorney with the Lathrop firm, and in the 1930s, John B. Gage was the "reform mayor" who overturned the Pendergast machine. Lathrop & Gage has produced four mayors, four state senators/representatives, one U.S. senator, one U.S. representative, three governors, and eight judges.

In recent years, the firm has championed many projects in the city's rebounding urban core, as well as across Missouri and Kansas. Lathrop & Gage participated in all aspects of the renovation of Union Station. It also negotiated the process through which the Internal Revenue Service achieved congressional authorization to lease the former Main Post Office; the new IRS service center is bringing six thousand jobs downtown.

As for new construction projects, Lathrop & Gage is providing counsel on the Metropolitan Kansas City Performing Arts Center and the National Association of Basketball Coaches' Hall of Fame, which will be adjacent to the Sprint Arena. Keeping with the history of the firm's first client, Lathrop & Gage attorneys and government relations advisors are working to develop a new intermodal facility for BNSF Railway Co. just outside Gardner, Kansas.

"It has been an honor and a privilege to work with so many individuals toward a single goal—to serve the Kansas City metropolitan area—by providing legal services to clients, leadership to community organizations, and manpower and funding to the philanthropic community," said Thomas S. Stewart, the firm's managing partner. ♦

> "IT HAS BEEN AN HONOR AND A PRIVILEGE TO WORK WITH SO MANY INDIVIDUALS TOWARD A SINGLE GOAL—TO SERVE THE KANSAS CITY METROPOLITAN AREA."

◀ Lathrop & Gage is proud to represent YRC Worldwide and its family of companies, including Yellow Transportation, Roadway, Reimer Express, USF, New Penn Motor Express, and Meridian IQ. Pictured: Michael Williams, Michelle Russell, Doug Dalgleish, Dan Churay, Bob McKinley, and Stacy Andreas.

PHOTO BY SCOTT INDERMAUR

At the Gould Evans office in Westport, the Corridor Art Space allows the firm to host exhibitions of work by local and regional artists. Gallery receptions serve as relaxed mixers for associates, clients, artists, and community residents.

Gould Evans associates gathered to help paint as part of renovating the headquarters of reStart Inc., an organization that assists in finding housing for thousands of individuals and families in the Kansas City area.

PHOTO BY ALAN S. WEIN

CHAPTER TWO: WORKING IN GREATER KANSAS CITY

The recent renovation of the Blue Springs South High School stadium and field has been a big hit with the school's Jaguars soccer team. Blue Springs schools have produced several state championships and titles in a number of different sports.

▲ *Blackwell Sanders Franchise & Distribution attorneys John Moore (left) and Don Culp learn the ins and outs of the frozen custard business from Jim Sheridan, president of Sheridan's Franchise Systems Inc.*

Blackwell Sanders Peper Martin LLP: Innovative Legal Solutions for a Complex Business World

As one of the leading commercial law firms in the Midwest, Blackwell Sanders Peper Martin excels at providing innovative legal strategies and solutions that meet the challenges of today's fast-paced global business environment. Its success is built upon a foundation of long-standing relationships both with middle-market public companies and some of the largest privately held companies in the United States.

"It has been our fortune to have attracted what we refer to as a blue-ribbon list of clients," says the firm's chief marketing officer, Lynn Snelgrove. "These are companies that are not only leaders in their industries in the Kansas City region, but also nationally and globally."

Founded in 1916, the firm has grown to include over 320 attorneys, about 160 in Kansas City alone. With more than forty practice concentrations, Blackwell Sanders has developed into a multidisciplinary, geographically diverse firm capable of meeting a wide range of legal and regulatory needs, no matter the industry or its location. Currently, the firm has offices in Kansas City, St. Louis, and Springfield, Missouri; Omaha and Lincoln, Nebraska; Overland Park, Kansas; Belleville, Illinois; Washington, D.C.; and London, England.

CHAPTER TWO: WORKING IN GREATER KANSAS CITY

Following a long tradition of hiring lawyers not only for their legal knowledge and academic excellence but also for their distinctive leadership and experiences, Blackwell Sanders is consistently recognized as one of the top law firms in the Midwest. For four consecutive years, *Corporate Board Member* magazine has ranked the firm as one of the Best Corporate Law Firms in Kansas City and St. Louis. Additionally, fifty-six of the firm's attorneys are currently listed in *The Best Lawyers in America* publication.

Anchored in a client-first philosophy and solution-driven approach, Blackwell Sanders is committed to achieving its clients' objectives practically and efficiently. By continually maintaining internal expertise while aggressively seeking opportunities for external growth, the firm has developed practice focus through its four major departments: Corporate, Litigation, Labor and Employment, and Real Estate.

Strong practice successes include the firm's public and private urban core development. Recent work includes a $536.4 million project to develop H&R Block's headquarters in downtown Kansas City; a combined $100 million in redevelopment of two historic downtown hotels, the President Hotel and Hotel Phillips; and a $100 million Plaza Colonnade development on the Country Club Plaza. Blackwell Sanders has earned practice distinctions in energy, health care, franchising, finance, retailing, and other industries.

Just as it reaches out to companies across the globe with its legal services, Blackwell Sanders also regularly extends a helping hand to its local communities.

"Not only is our legal profile very satisfying because we've been able to establish long-standing relationships with top clients, but our community leadership profile is equally rewarding," says Snelgrove. "We

(continued on page 148)

> BLACKWELL SANDERS IS COMMITTED TO ACHIEVING ITS CLIENTS' OBJECTIVES PRACTICALLY AND EFFICIENTLY.

PHOTO BY SCOTT INDERMAUR

◀ *Blackwell Sanders INROADS interns Jason Backstrom (left) and Mario Reyes collaborate on a research project involving international business. Founded in 1970, INROADS provides leadership and career training to qualifying minority students. An INROADS partner since 1999, Blackwell Sanders helps prepare its interns for law school and lays the groundwork for a successful legal career upon graduation.*

BLACKWELL SANDERS PEPER MARTIN LLP

◀ *Blackwell Sanders attorney Kim Jones (left) chats with Anne Bowman of UMB Bank and Sharyl Kennedy of Horizon Academy at the True North Plaza Lights Reception. The True North series provides professional development and networking opportunities for female executives.*

▼ *Blackwell Sanders senior counsel Kathy Bussing and her YouthFriend of eight years, Tiffany Lester, discuss a homework assignment. YouthFriends connects young people with adult volunteers who meet weekly for mentoring and encouragement.*

(continued from page 147)

actively encourage everyone at the firm to get involved."

As a result, its partners, associates, and staff contribute both time and money to dozens of local organizations and groups based on both individual and company-wide interests. Substantial, long-term support is also provided to the Nelson-Atkins Art Museum, the Kansas City Public Library, Kansas City Public Television, Legal Aid of Western Missouri, Legal Services for the Eastern District of Missouri, and Operation Breakthrough.

In 2004, the firm relocated its Kansas City office to the top floors of the newly constructed Plaza Colonnade, sharing the building with the Kansas City Public Library's Plaza branch, housed on the first floor. With more than ninety years in Kansas City, the move represents yet another way in which this global firm still cherishes its local, hometown roots. ♦

PHOTO BY ALAN S. WEINER

The Rosedale Memorial Arch was dedicated in 1923 as a tribute to the soldiers who served in World War I. Located in Rosedale, on the Kansas side, the monument was designed by resident John LeRoy Marshall and is patterned after the Arc de Triomphe in Paris.

At the close of The Chamber's most recent Governors' Summit, a Platform for Action —focusing on life sciences, animal health, education, and public transit—was signed. From left to right: Missouri Governor Matt Blunt; Summit Co-Convener Don Hall Jr.; Chamber Chairman John W. Bluford; Chamber President Pete Levi; Kansas Governor Kathleen Sebelius; and Summit Co-Convener Gary Forsee.

Polsinelli Shalton Flanigan Suelthaus PC: Practicing Law in a New Light

In 1972, James A. Polsinelli and two other young attorneys founded a small law firm in the historic Country Club District in Kansas City, Missouri. Specializing in representation for small and entrepreneurial businesses, the partners helped their clients seek opportunities for growth while navigating various competitive and regulatory obstacles.

Since then, Polsinelli Shalton Flanigan Suelthaus PC has grown to become one of the premier business and real estate law firms in the Midwest for business clients both large and small. With nearly three hundred attorneys and offices in Kansas City, St. Louis, Chicago, New York, Washington, D.C., Overland Park and Topeka, Kansas, and Edwardsville, Illinois, Polsinelli is also one of the fastest-growing firms in the country.

A first-generation law firm, it has never rested on its laurels, and continues to seek its own opportunities for growth.

"The first word that comes to my mind when thinking about the firm is energy," says Chairman and CEO W. Russell Welsh. "Through our commitment, time, and service, we've created an energetic environment that makes us a great place to work and an effective team for our clients."

▲ *Clients, colleagues, and friends of Polsinelli Shalton Flanigan Suelthaus gather each year for the annual Plaza Lights Party at the firm's offices on the Country Club Plaza in Kansas City. The event is an open-house celebration traditionally held the Thursday before the Plaza's lights are turned off for the season.*

CHAPTER TWO: WORKING IN GREATER KANSAS CITY

Whether returning a phone call quickly or hopping a flight at a moment's notice to assist in a negotiation, that energetic service is one of four core values that drive the firm. The other three include a focus on client needs, a commitment to utilizing fresh ideas when crafting solutions to those needs, and bright attorneys who deliver innovative services with exceptional skill, knowledge, and integrity.

Over the years, the firm has developed expertise in four core practice areas as well, areas that drive their recent growth and form the basis for their future. By focusing on business law, real estate and public law, financial services, and litigation, the firm provides innovative and savvy representation to businesses at every stage of their development.

To encourage the energetic spirit that is the hallmark of its practice, Polsinelli continually seeks to broaden its perspectives by attracting, retaining, and mentoring the best legal and professional talents available. With diversity comes new approaches to client issues, enhancement of the firm's overall culture, and vital connections to the communities in which the firm lives and works.

An active participant in a variety of diversity initiatives, Polsinelli is a charter signatory of the Kansas City Metropolitan Bar Association's Managing Partners' Diversity Initiative and regularly participates in diversity training workshops. In October 2006, the firm was awarded the Jackson County Bar Association's Pyramid of Diversity Award for outstanding commitment to promoting diversity, especially through its scholarships.

The firm further extends its commitment to the community via an active pro bono practice, which every year commits several thousand hours in legal and support services to the underserved.

"Not only are we an energetic firm," says Welsh, "we're a values-driven firm. Giving back helps those in need, and it allows us to expand our knowledge and stay involved with our communities. And that makes us better lawyers." ♦

> "NOT ONLY ARE WE AN ENERGETIC FIRM, WE'RE A VALUES-DRIVEN FIRM."

◀ Pictured in one of Polsinelli's litigation workrooms are (left to right) attorneys Judy Yi, associate; Cathy Dean, litigation department head; W. Russell Welsh, managing partner; and Dan Zmijewski, associate. The firm is known for its outstanding reputation in providing legal services for a wide range of industries and is a recognized leader in business law, financial services, real estate, and litigation.

Built in 1998 to help increase production, the Harley-Davidson Vehicle and Powertrain Operations in Kansas City employs nine hundred people in the production of some of the coolest bikes on the road. From raw material to final testing, the plant produces portions or completed versions of the Sportster, Dyna, VRSC, VRSCA V-Rod, and Revolution models. The plant is open for tours during the working day, allowing visitors to see activities ranging from welding to wheel assembly, painting to polish. There's also a gift shop on-site filled with Harley-Davidson accessories, apparel, and souvenirs.

CHAPTER TWO: WORKING IN GREATER KANSAS CITY

It's hard to believe that heat—extreme heat—and the talent of an experienced glassblower like Dierk Van Keppel can transform ordinary compounds like sand, limestone, soda ash, and potash into spectacularly beautiful works of art. Van Keppel works in his studio, Rock Cottage Glassworks, using centrifugal force and gravity to blow his glass into art objects. Years of training and outstanding talent go into this ancient craft. As a traditional artisan, Van Keppel uses blowpipes, hand tools, and heat to produce his work. Liquid glass pushing two thousand degrees is gathered on steel blowpipes and controlled with wooden blocks and wet newspapers. Van Keppel's works fall into four general categories: architectural, lighting, art glass, and sculpture.

PHOTO BY MARIO MORGADO

PHOTO BY MARIO MORGADO

PHOTO BY MARIO MORGADO

Bank Midwest Invests in Kansas City's Future

Since its founding in 1970, Bank Midwest has become one of the largest and most successful privately held financial institutions in the region. This success has been due in large part to the bank's owners and executives continually looking to the future and being prepared with banking products and services to meet evolving customer needs.

Bank Midwest has continually expanded its consumer banking operations, entering new markets and opening new banking centers every year. The bank currently has more than thirty facilities throughout Greater Kansas City and will continue to open new locations to better serve its expanding customer base. Bank Midwest is unique among banks in the region in providing complete banking services via convenient in-store banks in over twenty Wal-Mart locations throughout the region. These banks also offer expanded hours of operation and discounted convenience services.

While consumer banking remains a high priority for Bank Midwest, its greatest growth has come in the commercial banking arena. Bank Midwest is committed

▲ *Pictured in the rotunda of Town Pavilion, a Bank Midwest–financed office tower and the bank's future headquarters, is the commercial real estate lending management group (left to right): John Baxter, senior vice president; Randy Nay, senior executive vice president; Rick Smalley, president and chief executive officer; Dan Dickinson, executive vice president; and John Price, senior vice president.*

CHAPTER TWO: WORKING IN GREATER KANSAS CITY

PHOTO BY ERIC FRANCIS

to investing in the future of Kansas City by aggressively pursuing opportunities to fund commercial development. Developers turn to the bank because their expertise and capital strength make them an ideal partner for bringing large and complex projects to life.

In addition to commercial real estate financing, Bank Midwest has developed expertise in financing for businesses in virtually all of the region's major industries, providing all types of commercial financing, including funding for the purchase of inventory and equipment, working capital, and accounts receivable.

While it has earned a reputation in its hometown as a top-tier resource for commercial lending, it is also known nationwide for its expertise in commercial real estate development financing. Developers and builders from coast to coast call on Bank Midwest to take advantage of its broad experience in designing credit facilities for

(continued on page 156)

THE FUTURE COMES ONE DAY AT A TIME, AND BANK MIDWEST'S COMMITMENT TO THE FUTURE OF KANSAS CITY IS EVIDENT IN THE WAY IT DOES BUSINESS EVERY SINGLE DAY.

◄ Seated in the Harry S. Truman rail car at the headquarters of good customer Kansas City Southern are (left to right) Bank Midwest officers Mark McCaskill, vice president; Bobbie McCauley, senior vice president; Quintin Ostrom, senior vice president; and Robert Parks, senior vice president. Standing are (left to right) Damon Stelting, vice president; Paul Holewinski, executive vice president; Dave Rambo, senior vice president; Charlie Koch, vice president; and Brian Bower, vice president.

PHOTO BY ERIC FRANCIS

BANK MIDWEST | 155

BANK MIDWEST

◀ Pictured in front of a suburban Bank Midwest banking center are (left to right) retail banking officers Darren Bell, senior vice president; Tim Thomas, executive vice president; Steve Mallory, executive vice president; Mike Sinnett, senior vice president; Sue Greenway, senior vice president; Gerry Clemen, senior vice president; and Bill Arnold, senior vice president.

(continued from page 155)

income-producing residential properties, investment properties, owner-occupied commercial real estate acquisition and construction financing, and sophisticated mezzanine and equity sponsor financing.

Owners of small businesses also find an extensive lineup of services at Bank Midwest. The bank's Small Business Advantage product package provides advanced treasury management and depository services on par with those once reserved for only the largest businesses. This includes secure online banking and advanced treasury services like remote deposit capture, image-based lockbox services, payroll cards, and more. The Small Business Advantage also includes credit facilities that are available through a streamlined application process.

Bank Midwest is committed to the future of Kansas City in ways that extend beyond serving its banking needs. The bank's employees are actively involved in numerous community organizations and activities that benefit those less fortunate residents. The bank's owners and executives also serve on the boards of numerous civic organizations, and contribute generously to a long list of worthy causes that are crucial to the long-term growth and development of the city.

The future comes one day at a time, and Bank Midwest's commitment to the future of Kansas City is evident in the way it does business every single day. ♦

▼ Pictured in front of the Bank Midwest in-store banking center inside a metro area Wal-Mart Supercenter are banking center manager Nate Boylan (left) and regional manager Gerry Clemen, senior vice president.

CHAPTER TWO: WORKING IN GREATER KANSAS CITY

Wireless may be all the rage for many electronic functions, but electricity still powers the world. And human beings still make sure it flows efficiently. Each year since 1984, the International Lineman's Rodeo Association, headquartered in Kansas City, Missouri, has honored this important and often dangerous work with a competition held in Bonner Springs, Kansas. The rodeo tests the skills and techniques of linemen from around the world, and in 2006, over 200 teams and 250 individual apprentices competed in specialized rescue techniques, pole climbing, and a series of mystery events. The association also holds a yearly expo as well as a training and safety conference, and funds a scholarship program that promotes the lineman trade by assisting students attending electrical lineman training school.

Black & Veatch Corporation: Making a Global Impact from the Midwest

What do Singapore, Hong Kong, Scottsdale, Arizona; and Milwaukee, Wisconsin, all have in common? They each have award-winning infrastructure projects that reflect the talent and expertise of Kansas City–based professionals from Black & Veatch, the global engineering, consulting, and construction company serving clients since 1915.

"The intellectual capital we apply to projects around the world has resulted in many awards from industry associations and clients, recognizing both our innovations and technological breakthroughs that we provide to our clients' most complex challenges," said Len Rodman, chairman, president, and CEO of Black & Veatch. "Our global reach also means we have global resources available for any project, regardless of location. While our roots are local, our impact is most definitely global."

The company expresses that desired impact in its mission statement, *Building a World of Difference*®, which means that each Black & Veatch professional has the opportunity to improve the living standards of millions of people. Whether it's

▲ *The global nature of Black & Veatch is reflected in its diverse workforce, with about seventy languages spoken throughout the Kansas City locations.*

CHAPTER TWO: WORKING IN GREATER KANSAS CITY

electricity at the flip of a switch, clean and safe water from the tap, or instant telecommunications in an emergency, Black & Veatch provides it seamlessly every day.

With more than one hundred offices across the globe, the nearly nine thousand worldwide professionals at Black & Veatch have solved complex engineering and construction problems for clients on six continents in more than one hundred countries. In Kansas City, thirty-six hundred professionals often collaborate with coworkers across multiple countries and time zones, resulting in projects that can be worked on literally twenty-four hours a day.

Black & Veatch has earned the number-one ranking in power design by the respected *Engineering News-Record* magazine, and ranked number two in power transmission and distribution design. These rankings are not surprising, since the company has been involved in more megawatts of power generation than any other company in the world.

Providing leading technology in coal gasification for its clients, Black & Veatch is also currently involved in a large percentage of the coal-fired power plant projects across the United States.

Black & Veatch is an industry leader in sulfur recovery, including expansion work at the world's largest oil refinery in Jamnagar, India. The company's air quality unit has nearly one-third of the market share of the vital clean emissions projects at U.S. power plants, and is the licensee for the leading clean coal "scrubber" technology. Black & Veatch has designed and built more gas turbine facilities than any company in the world, and has experience in advanced nuclear plant design/build projects.

(continued on page 160)

"WHILE OUR ROOTS ARE LOCAL, OUR IMPACT IS MOST DEFINITELY GLOBAL."

◀ *Black & Veatch is providing engineering, procurement, and construction services at the Nebraska City Station Unit 2. The new unit will provide electricity to three hundred thousand Omaha Public Power District customers and seven other public power entities.*

BLACK & VEATCH CORPORATION | 159

BLACK & VEATCH CORPORATION

◀ In Tian'e, China, the world's largest roller-compacted concrete dam will ensure that 12 million people in China have reliable electricity and protection from flooding.

(continued from page 159)

The company's impact in water supply and water treatment is realized by millions across the globe. In fact, an estimated 20 percent of the world's population drinks water through systems designed, constructed, or supported by Black & Veatch. In the Kansas City metropolitan area, that figure jumps to about 90 percent for water supply, due to the company's work with the Kansas City Missouri Water Services Department, Johnson County WaterOne, and others. In addition, 70 percent of the metro area uses wastewater systems that involve Black & Veatch expertise.

Awards just seem to naturally follow many of Black & Veatch's water projects. The space-saving design at the Tai Po Water Treatment Works in Hong Kong prompted the Global Grand Prize in Design Award from the International Water Association. At the 2006 Global Water Awards, the Singapore-Tuas Desalination project—the largest of its kind in Asia—was given the Desalination Plant of the Year Excellence Award. The Milwaukee Northwest Side Relief Sewer project received the Project of the Year Award from the American Public Works Association. In Scottsdale, Arizona, the

◀ Black & Veatch has delivered telecommunications solutions on more than twenty-seven thousand wireless cell sites, resulting in more than 5 million traffic hours daily.

The space-saving design at the Tai Po Water Treatment Works in Hong Kong prompted the Global Grand Prize in Design Award from the International Water Association.

new Chaparral Water Treatment Plant garnered two awards: the Grand Award in Engineering Excellence and the Art in Public Places Award.

Through its telecommunications work, Black & Veatch has deployed the first U.S. fiber-optic and fiber-coax networks, first PCS network, first satellite entertainment network, and first multimedia network for cellular handsets. Locally, Black & Veatch has been instrumental in North Kansas City's new community-wide fiber-optic network, providing Internet connectivity to twenty-five hundred homes and nearly one thousand businesses.

The company has also been providing services for the U.S. federal government sector since World War I. Projects include infrastructure rehabilitation in Afghanistan and Iraq, and disease monitoring work in Russia, which has been useful for disaster preparedness being carried out locally in Johnson County, Kansas.

"Our work is not always in the spotlight," says Rodman, "but we know that people depend on our systems every day to meet their energy, water, and telecommunications needs in a manner that is reliable, innovative, and sustainable." ♦

The Black & Veatch–designed cooling tower at the Nearman Creek Power Station provides a backup cooling source to ensure a reliable supply of electricity to customers of the Board of Public Utilities in Kansas City, Kansas.

Known as the "Heart of America"—the maximum distance to anywhere in the continental United States is approximately nineteen hundred miles—Kansas City is equally beautiful day or night. The skyline is an interesting blend of art deco buildings and modern skyscrapers. In the mid-1990s, the skyline took on a dramatic new look with the installation of four massive steel sculptures that rise more than two hundred feet above the city. The sculptures, *Sky Stations/Pylon Caps*, are lighted at night and can be seen for miles.

CHAPTER TWO: WORKING IN GREATER KANSAS CITY

Fogel-Anderson Construction Co.: Four Generations of Building the Best

When 70 percent of your company's clients are repeat or referral customers, the *Kansas City Business Journal* ranks you as one of the best contractors in the area, and you've been in the building business for over ninety years, you've certainly earned the right to use "Build with the Best!" as your slogan. Fogel-Anderson Construction Co. was incorporated in 1917 and has since remained family-owned. Douglas W. Fogel, executive vice president, said, "We have a tremendous group of clients and employees, and we like to say that we build better projects one brick at a time."

As specialists in the commercial and industrial construction market, Fogel-Anderson is a preferred regional contractor for the Lowe's Home Improvement Company stores and is strongly considered for construction of distribution centers. They are a primary regional builder of Wal-Mart Supercenters, and the local electric utility, Kansas City Power and Light, has been a client since before World War II. Fogel-Anderson recently contracted its ninety-fifth job with the same client—now that's a repeat customer! The company provides professional services to developers, retailers, hotels, theaters, and banks, and office buildings for individual business owners.

Fogel-Anderson guarantees a "service edge" to every single client, a phrase that can be defined with two words—*personal involvement*. "From the CEO to our field

▲ *Northridge Plaza in Olathe, Kansas, is a twenty-five-acre development including site work and a power center. The development features such popular stores as Old Navy, Dick's Sporting Goods, and Bally's Fitness Center. Outlots spotlight family restaurants Chipotle, Panera Bread Co., Ruby Tuesday's, and IHOP.*

personnel, all of us examine every project and decide the best way to successfully complete it," said Fogel. Attention to detail and enthusiastic customer service require happy employees. This firm views client satisfaction as a primary goal and constantly reinforces the value of mutual support. The Fogel-Anderson environment results in customers that receive top service and employees who remain at the company for decades: Ted A. Anderson, chairman of the board and CEO, has been with the company for fifty-five years and to this day spends time working one-on-one with clients. Phillip D. Bartolotta, president and COO, has hands-on status in every facet of the company's operations.

Whether in the role of general contractor, design builder, or construction manager (overseeing the entire project for an owner), Fogel Anderson delivers competitive cost control, maximum quality, and on-time completion. Clients can also count on safety. The organization has repeatedly earned the Platinum-Level STEP (Safety Training and Evaluation Process) Award, a national recognition earned by few, because of its stringent guidelines for safe working practices and conditions.

"We are a company that applies our resources wholeheartedly, ethically, and honestly for our clients," said Fogel. The firm proudly supports over thirty-six local charities via auctions, tournaments, and direct donations. Community involvement adds to the satisfaction of doing a good job every day, but the real enjoyment comes from exceeding clients' expectations. Fogel explained, "If you want to work in an industry where you are challenged to be beyond your best on a daily basis, this is the business to be in." ♦

"WE LIKE TO SAY THAT WE BUILD BETTER PROJECTS ONE BRICK AT A TIME."

◀ Fogel-Anderson Construction Co. served as the general contractor for the Blue Ridge Bank and Trust Co. corporate offices in Independence, Missouri, completing the project in sixteen months. The six-story, 113,000-square-foot structure has a reinforced concrete frame and aluminum curtain walls, and stands 113 feet high.

Chef Bob Brassard (far left) provides culinary arts instruction at Broadmoor Technical Center. As part of instruction in the culinary programs, students get to practice their skills each month by preparing a multiple-course meal for public consumption, with tickets sold to guests from around the community who come to savor the cuisine. A signature program of the Shawnee Mission School District, in Shawnee Mission, Kansas, Broadmoor Technical Center offers morning and afternoon sessions that give students from area high schools a chance to learn employable skills in specific areas of industry. From culinary arts to small engine repair, Broadmoor offers hands-on education that prepares students for the working environment. Students attending Broadmoor's programs have an opportunity to earn a technical certification or be awarded scholarships, and a majority of those taking part in Broadmoor's offerings pursue postsecondary education upon graduation from high school.

CHAPTER TWO: WORKING IN GREATER KANSAS CITY

Carl Sagan once said, "If you want to make an apple pie from scratch, you must first create the universe." Marcia Prentiss, owner of the Pie Lady Coffee House, doesn't go that far, but she does know the time and effort it takes to make a dream a reality. Opened in December 2001, her café is the culmination of years of selling her homemade pies at local farmers markets—where she became known as the "pie lady"—and a vague dream of one day opening her own pie shop. In 2000, she took an entrepreneurial course at Johnson County Community College and a year later was in business. Located in historic downtown Lenexa, the Pie Lady Coffee House specializes in daily made-from-scratch pies, everything from meringue and cream to pecan and fruit pies featuring only fresh, seasonal fruits. Apple pie is a best seller year-round, and in early summer, people start lining up for the strawberry/rhubarb.

CHAPTER TWO: WORKING IN GREATER KANSAS CITY

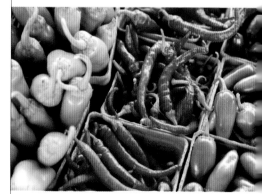

Imagine an open-air market filled with an array of products, shops, and the sound of many different languages being spoken. Are you in a European village? No, you're in the City Market, in Kansas City, Missouri. The market, open seven days a week, provides wagon loads of meats, specialty foods, flowers, and gifts from around the world. Originating in 1857 as a site for horse trading, political rallies, medicine shows, and circuses, the City Market became a hub for pioneers stocking up on supplies before heading westward. Today the market continues to thrive and is also Kansas City's third-largest outdoor concert venue. The Farmers Market is available every weekend from mid-March to November, packed with fresh fruits and vegetables, baked goods, and crafts from area artists and farmers. Here, John Carollo cooks up some barbecue; Haiyan and David Lee sell flowers; and Phillip and Esther Montelgano peruse the produce.

Gould Evans: Creating Spaces That Enrich Life

From its start in 1974, Gould Evans has been a firm that places great value on innovative thinking and collaboration as central to the design process. Commitment to excellence and collective effort have driven the firm's success at each of its seven offices around the country, including the three in the Greater Kansas City region: one in the Westport neighborhood in Kansas City, Missouri; one in Overland Park; and one in Lawrence, Kansas.

"We have always sought to nurture strong client relationships with a collaborative process," explains principal Dennis Strait. "This has helped us find success with all types of clients. They come to us for a particular expertise we offer, such as design or master planning, then become involved with our inclusive approach and find outcomes that support their missions and programs."

To explore how people will live, learn, work, and play in space and to design solutions that are beautiful and highly effective, the firm brings together teams of design and planning professionals with clients in collaborative workshops. These collective explorations allow the teams to address all aspects of a project: its immediate

▲ *Gould Evans designed an addition and renovation of the Harry S. Truman Presidential Museum and Library. The facility's educational spaces and exhibits include a primary gallery with a strong visual connection to the gravesite and outdoors.*

CHAPTER TWO: WORKING IN GREATER KANSAS CITY

PHOTO BY MIKE SINCLAIR

The Gould Evans portfolio includes projects that vary widely in size and purpose, from cultural and civic structures to residences and retail centers to educational facilities and institutes of technology.

"A healthy firm is diversified," says principal Scott Stalcup. "This belief has encouraged us to work with both public and private clients. Our associates enjoy the diversity of work, and the work itself benefits from the diverse knowledge base within the firm." Gould Evans has cultivated strong relationships with a broad array of educational institutions, government agencies, private corporations, and nonprofit groups.

Gould Evans leaders see the firm as a learning organization that is committed to the future of the architecture and allied professions, which is one reason the firm has implemented a mentoring program that helps prepare the next generation while at the same time reinvigorating working

(continued on page 172)

PHOTO BY MICHAEL SPILLERS

"THE ARCHITECTURE PROFESSION HAS A RESPONSIBILITY TO ADDRESS THE ISSUE OF PLACE IN A HOLISTIC WAY."

◀ *The Community Health Facility in Lawrence serves three health organizations and several patient populations in one building. A strong civic profile and a public art component help define it as a facility open to the public.*

and future functions, issues of resource use, responsiveness to bioclimate, impact and connections to the surrounding community, and more.

"The architecture profession has a responsibility to address the issue of place in a holistic way," says principal Tony Rohr of the process. "This is one reason that we are a multidisciplinary firm. We believe that cross-disciplinary thinking enriches the whole team and every project. And it relates to the important task of thinking simultaneously about different scales. Thinking about the street, the neighborhood, and the region is a critical imperative to creating a successful project and community."

GOULD EVANS | 171

GOULD EVANS

PHOTO BY MIKE SINCLAIR

◀ *The Legends 14 Theatre at Village West in Kansas City, Kansas, is an exemplary public/private partnership aimed at community investment and revitalization. The superstructure at the entry provides an iconic marker in the outdoor courtyard.*

(continued from page 171)

professionals. "We started the intern development and mentoring programs as a way to help support our young people on the path to licensure," says principal Becky Mullins. "In the process, we created a mentoring program that has value to that group and to those with more experience. It's a two-way structure. The program has allowed associates to actively participate in community service projects, supporting an overall firm vision of community embeddedness."

That vision is further evidenced by the firm's Westport office location, an expression of the firm's commitment to the urban core and to investment in existing and historic neighborhoods. "We believe that healthy cities include vibrant, diverse urban neighborhoods, and we have worked hard to be a good neighbor in ours as well as continually seeking out projects that enrich the fabric of our community," says Rohr.

Principal Bob Gould adds that the firm's loyalty to its cities goes beyond the working day. "What we do is about place, and perhaps because of that, it has always

▼ *In Overland Park, Johnson County Community College's new Regnier Center for Business and Technology will be part of a complex that also includes the Nerman Museum of Contemporary Art. The center is designed to represent and support the intersection of human ingenuity and technology.*

RENDERING BY ARNOLD IMAGING

CHAPTER TWO: WORKING IN GREATER KANSAS CITY

◀ The Cerner Corporation campus accommodates thousands of knowledge workers and is designed to be a visual and physical representation of the company's attributes, which encompass technological advancements for health, innovation, and leadership.

▼ The Fellowship Hall addition to the Grace and Holy Trinity Cathedral in downtown Kansas City helped extend the ministry and mission of the church. This project was a collaboration with Taylor MacDougall Burns.

PHOTO BY MICHAEL SPILLERS

PHOTO BY TIMOTHY HURSLEY

been part of our firm's philosophy to be rooted to our own communities." This approach includes support for local philanthropic organizations, many of which benefit from active involvement of Gould Evans associates as well. Collectively, they are involved in many groups. The spirit of giving among associates is an energy matched only by their desire to create places that are pleasurable and purposeful. ♦

GOULD EVANS | 173

PHOTO BY ALAN S. WEINER

174 | GREATER KANSAS CITY: UNLIMITED POSSIBILITIES

CHAPTER TWO: WORKING IN GREATER KANSAS CITY

PHOTO BY ALAN S. WEINER

The Zona Rosa District in Kansas City's Northland literally has everything from A to Z—Abercrombie & Fitch to Zales—and all the stops in between. The area around Terrace, Prairie View Road, Stoddard Avenue, and Dixon offers places to live, an extensive range of restaurants serving every kind of food imaginable, an equally extensive selection of shops, plus spas, entertainment, beauty salons, and art galleries. There's even a dentist available. This is Kansas City's newest and hottest destination for both adults and kids. Without a doubt, there's something for everyone.

Lewis, Rice & Fingersh: National Recognition. Personal Service.

Lewis, Rice & Fingersh represents a wide spectrum of clients locally, regionally, and nationally by providing a comprehensive array of legal services designed to meet each client's needs.

"We believe we're very connected to our clients," states William E. Carr, managing member of the Kansas City office. "Quality work and long-term client relationships are the foundation of everything that we do.

"Our attorneys are responsive problem solvers. They understand the challenges that our clients face every day and utilize the right experience, talent, and resources to help our clients accomplish their business goals."

The level of personalized service and continuity maintained by Lewis, Rice & Fingersh may be unique in today's often changing legal landscape, and has led to some decades-long professional relationships.

Lewis, Rice & Fingersh is a firm of some 170 attorneys with seven offices in Missouri, Kansas, and Illinois. The firm's reputation for legal excellence is based upon a foundation started individually and strengthened by the merger of Lewis & Rice (begun in St. Louis in 1909) and Brown, Koralchik & Fingersh (founded in Kansas City in 1948) in 1989 to form what Carr refers to as the "right size" firm—large enough to provide full service and to be responsive to clients' needs, while still small enough to allow for personal relationships.

▲ *(Left to right): Stan Johnston, chairman, Business, Tax & Estate Planning; Bill Carr, managing member; Charles Miller, chairman, Real Estate; and Robert Tormohlen, chairman, Litigation, gathering for their monthly executive committee meeting.*

CHAPTER TWO: WORKING IN GREATER KANSAS CITY

PHOTO BY SCOTT INDERMAUR

From its Kansas City office, Lewis, Rice & Fingersh stays especially attuned to the real estate, corporate, estate planning, and litigation needs of business clients.

In real estate, the firm's expertise touches all aspects of deals—acquisitions and sales, zoning, title work, public and private financing, leasing, and the full scope of taxation matters. The practice covers projects nationwide from residential subdivisions and condominiums to shopping centers, office complexes, industrial parks, and mixed-use developments.

In corporate law, representation encompasses closely held private companies, handling concerns ranging from mergers and acquisitions to employment relations, as well as start-ups. Attorneys in this practice area also specialize in estate planning, wealth management, and trusts.

And in matters involving litigation, the firm navigates a diversity of issues including general business litigation, as well as insurance, banking, securities, employment, and environmental law. The group strives for excellent results without protracted litigation and related costs.

For clients of Lewis, Rice & Fingersh, representation is about both close associations and quality representation. Carr summarizes the firm's success as follows: "People often say it's all about building relationships, but maintaining those relationships stems from being good at what we do." ♦

"QUALITY WORK AND LONG-TERM CLIENT RELATIONSHIPS ARE THE FOUNDATION OF EVERYTHING THAT WE DO."

◀ *Anna Blevins, paralegal, joins Carlos Lewis, associate, and Peter Hartweger, member, in reviewing a client agreement. The three work in the firm's Business, Tax & Estate Planning department.*

PHOTO BY SCOTT INDERMAUR

LEWIS, RICE & FINGERSH, L.C. | 177

"Anyone who doubts the relevancy of classical music in today's pop musical culture need only go to the movies," says violinmaker Anton Krutz. "From *Psycho* to *Jaws* to *Star Wars*, if you take out the music, you lose much of the emotional impact." Perhaps that is because, he says, string instruments come closest to mimicking the human voice. He should know. Shown here in the maroon shirt, Anton has been making violins since he was twelve years old and is a graduate of the internationally known Violin Making School of America in Salt Lake City. In 1992, he partnered with his father and Kansas City Symphony bassist, Misha Krutz, and bass-maker Rick Williams (blue shirt) to open K.C. Strings Violin Shop in Merriam, Kansas, which specializes in making and repairing violins, violas, cellos, and basses for clients around the world. The most pleasurable aspect of his craft is, says Anton, "creating instruments that have a beautiful sound that lives on for centuries afterward."

CHAPTER TWO: WORKING IN GREATER KANSAS CITY

Throwing a pot is easy . . . if you have a good teacher. In this case that teacher is Rebecca Kopp (in green), who is helping a student get the hang of it. Rebecca opened Back Door Pottery in 1979 and moved to her present location on St. John in 1986. All her work is done here in the studio, and there is a small showroom in front. Classes are held two evenings each week and are open to beginners as well as advanced recreational potters. Rebecca also conducts children's workshops and individual sessions on an appointment basis. Many residents know her work—mostly functional, such as kitchenware—which can be found in a number of specialty shops and galleries around Kansas City.

GREATER KANSAS CITY: UNLIMITED POSSIBILITIES | 179

JE Dunn Equals Honesty and Integrity

"The only things we build better than our buildings are our relationships," says William H. Dunn Sr., chairman emeritus of JE Dunn Construction. This family-owned company, founded in 1924 by William's father, John Ernest Dunn, has grown to be the sixth-largest general building contractor in the country.

"We're certainly proud of that accomplishment," says Terrence Dunn, the third-generation president and CEO, "but throughout that time, we've maintained our family values and we're equally proud of that."

A prime example of the importance of those core values is that JE Dunn gives 10 percent of pretax profits to charities. In addition, employee involvement in community is encouraged. "We've created an environment for our employees here in Kansas City, as in all our locations, that encourages them to get involved in the communities in which they live and work." A long list of Dunn associates sit on local boards, and the company supports more than four hundred nonprofit organizations nationwide.

The philanthropic environment is diffused throughout the organization and has

▲ *With the help of JE Dunn, the Nelson-Atkins Museum of Art underwent its first expansion since 1933. The new Bloch Building, illuminated in the background, expanded the total museum space by 60 percent. The water feature,* One Sun/34 Moons, *is in the foreground and caps a 457-space underground garage.*

CHAPTER TWO: WORKING IN GREATER KANSAS CITY

PHOTO BY THOMAS S. ENGLAND

been since Bill Dunn Sr. took sole control of the business in 1974. Dunn states it very simply, "It's important to leave the world in better shape than when you came into it."

In keeping with the standards of the founder, Terry Dunn is quick to point out, "Although our family has the ownership, it's our employees who have the leadership. We can provide the direction, but it's the folks who come to work every day that make this company what it is."

Barry Brady was a JE Dunn employee for twenty years before leaving to manage real estate holdings throughout Kansas City. He now sits on the JE Dunn board of directors. "It's amazing that any business would succeed for now a third and fourth generation. It's a testimony not only to the Dunns but also the team they've assembled."

Specifically, JE Dunn is a $2.6 billion company with plans to continue to grow that figure in future years. With seventeen offices in the United States, the JE Dunn Construction Group is the holding company for six independently run construction companies. Add to that an insurance company, an equipment company, and real estate investments. "We're diversified, but basically we're centered around construction," Terry Dunn explains.

From building to redevelopment, JE Dunn's involvement in Kansas City can be seen in the skyline, with buildings like the six-hundred-thousand-square-foot Federal Reserve Bank on fifteen acres downtown; the seventeen-story H&R Block World Headquarters office tower atop a five-level underground parking area; the massive IRS Processing Building; the innovative expansion of the Nelson-Atkins Museum of Art; the stunning restoration of the President Hotel; the breathtaking plans for the

(continued on page 182)

"IT'S IMPORTANT TO LEAVE THE WORLD IN BETTER SHAPE THAN WHEN YOU CAME INTO IT."

▼ *The mission of JE Dunn is "to provide construction services in a professional manner to exceed the expectations of our clients." The success of this formula is evident throughout the Kansas City area in buildings such as the University Academy.*

JE DUNN CONSTRUCTION

PHOTO BY THOMAS S. ENGLAND

◀ Chances are that this is the most interesting parking garage you'll ever see. JE Dunn was instrumental in this and other projects to revitalize downtown. Citizens were asked to pick influential books that represented Kansas City. The titles became "book-bindings" for the exterior of the Downtown Central Library parking garage.

(continued from page 181)

Kauffman Center for the Performing Arts; and the 240-acre Sprint campus with twenty-two buildings and sixteen parking garages—plus 60 percent green space—in the Kansas City suburb of Overland Park.

"JE Dunn has had a tremendous impact on the community from the standpoint of a general contractor and a construction manager. In other words, a good bit of our skyline has Dunn's fingerprints on it," says Hugh Zimmer, chairman and CEO, Zimmer Companies.

Another area of accomplishment is health-care construction. A story unto itself, JE Dunn began building health-care facilities in the 1930s and '40s. The company has been involved in the building, expansion, or renovation of most of the major hospitals in the Kansas City area. Moreover, the company's current advance into lifescience/bioscience construction is particularly exciting.

"This is highly complex construction, and not all firms are qualified," Dunn explains. "To serve this market, we've created the Healthcare Center of Excellence, whose mission is to maintain a trained,

PHOTO BY ALAN S. WEINER

◀ The Command and General Staff College at Fort Leavenworth, Kansas, is one of the nation's most technologically advanced learning facilities. Full technological integration allows students to use data, voice, and video in a variety of learning simulations.

CHAPTER TWO: WORKING IN GREATER KANSAS CITY

◀ *The IRS Processing Building consolidates IRS processing operations under one roof. During peak tax season, the new building will accommodate over six thousand employees. The building received a GSA Design Excellence Honor in Lease Construction.*

in-house staff to keep up with the latest trends in technology, equipment, and research systems.

"We believe the combination of having the ability to tap into the strength of our national construction company, plus the specialized knowledge provided by our Healthcare Center of Excellence, enables us to provide our high-tech health-care customers with the best service possible."

JE Dunn's future mission, as it has been in the past, will continue to be the best client-centered building partner, providing construction services in a professional and highly ethical manner, and exceeding its clients' expectations. ♦

PHOTO BY THOMAS S. ENGLAND

◀ *Since the early 1990s, JE Dunn Construction has had a lasting relationship with Children's Mercy Hospital. The company has completed expansions, additions, and remodeling projects, including enlarging the inpatient tower facility and clinic space, and adding a four-hundred-car parking structure.*

CHAPTER TWO: WORKING IN GREATER KANSAS CITY

PHOTO BY MARIO MORGADO

If the Union Station clock could talk, it would tell tales dating back to 1914. Since that time, people have been connecting under the clock for business meetings, family trips, engagements, and more. Restored to its original splendor, Union Station is one of Kansas City's most distinctive landmarks. Primarily a destination for a wide variety of entertainment and cultural events, the station is also home to Science City, an interactive museum for kids to experience hands-on educational fun. Rail travelers can catch any one of three AMTRAK rail lines from the station: the Southwest Chief, the Ann Rutledge, and the Missouri Mule. In addition to active trains, the station is home to a permanent exhibit of vintage rail cars, period artifacts, N-gauge model trains, and storytelling ghosts who share tales of travel in days gone by. The station also houses five eating outlets for hungry visitors or anyone nearby who needs a light snack.

PHOTO BY MARIO MORGADO

Armstrong Teasdale: Providing Invaluable Resources to a Diverse Family of Clients

From Kansas City to Shanghai, Washington, D.C., to San Francisco, Armstrong Teasdale has evolved from a two-attorney operation to more than 265 attorneys practicing in eleven locations. For more than one hundred years, Armstrong Teasdale has been providing invaluable legal resources and practical advice to a diverse family of national and international clients requiring litigation and corporate law services.

The Kansas City office was established in 1883 by John D. S. Cook and Alfred Gossett. These two men practiced together until 1907, when Cook retired. A century later, the firm was named Dietrich, Davis, Dicus, Rowlands, Schmitt & Gorman,

▲ *Members of Armstrong Teasdale's Explosion, Fire and Electrocution Practice Group stand in front of a statue in downtown Kansas City of Ike Davis, former mayor of Kansas City and managing partner of the firm's Kansas City office.*

and was reported to be one of Kansas City's oldest law firms. In 1988, St. Louis–based Armstrong, Teasdale, Kramer, Vaughn and Schlafly and Dietrich Davis initiated preliminary merger discussions. Later that year, formal negotiations began, and the merger became effective on January 1, 1989, under the new name of Armstrong, Teasdale, Schlafly, Davis & Dicus. The name was later shortened to Armstrong Teasdale in 1999. Ike Davis, former two-term mayor of Kansas City and former managing partner of the Kansas City office, is among the distinguished attorneys to have practiced at the firm.

Armstrong Teasdale has established a solid reputation of excellence and serves a dynamic national and international client base in virtually every area of law. Offering a full range of innovative legal services, the Kansas City office primarily focuses on business and tort litigation, franchise and distribution, public law and finance, white-collar criminal defense, real estate, and employment and labor. One of the firm's fastest-growing practice areas is the highly specialized explosion, fire, and electrocution practice. This Kansas City–based group is involved in the day-to-day management and trial of explosion, fire, and electrocution cases on a national basis. Attorneys within the practice group are leaders in the industry and are committed to the continued advancement of the practice, by actively participating in professional and legal organizations that influence the areas of science and law.

Armstrong Teasdale and its attorneys have recently received national recognition, such as the selection of sixty-two attorneys for inclusion in the thirteenth edition of *The Best Lawyers in America* (2007) and the acknowledgment of outstanding client service by BTI Consulting Group, which surveyed more than two hundred Fortune 1000 companies to ascertain which law firms provide the best services.

"We strive to be valuable partners when relating to a client's business in an effort to protect their best interests," explains Lynn Hursh, managing partner of the Kansas City office. "Clients are regularly kept apprised of the status of their cases, and participate with us in decisions concerning their matters. It's more of a collaborative effort than the traditional client/attorney relationship."

Armstrong Teasdale is committed to providing the highest-quality legal services to Kansas City's local business and civic community today and in the future. ♦

> "WE STRIVE TO BE VALUABLE PARTNERS WHEN RELATING TO A CLIENT'S BUSINESS IN AN EFFORT TO PROTECT THEIR BEST INTERESTS."

PHOTO BY MARK McDONALD

◀ *(Left to right) Kansas City Real Estate Practice group leader Randal J. Leimer and Kansas City managing partner Lynn W. Hursh, who represent a firm with a solid reputation for excellence that serves clients around the globe in virtually every area of law.*

Joe Polo, owner, and Nina Ward, co-founder, of Original Juan Specialty Foods are seen here next to Juanita, the icon of the restaurant chain that spawned the idea for the food company. Founded in 1998 as a manufacturer of hot sauces, salsa, and other specialty foods, Original Juan outgrew its first location in only six years and moved just up the road to its present digs at 111 Southwest Boulevard. The company manufactures 150 items right on the premises, ranging from sauces and seasonings to dressings and snacks. In addition to supplying area restaurants and retailers, and online sales throughout the continental United States, Original Juan sells direct to the public through its factory outlet. While its products and business practices have brought the company numerous awards, its owners say the real reason for its success is adherence to an ingrained set of quality values and a belief in living life to the fullest beyond the working day as well.

CHAPTER TWO: WORKING IN GREATER KANSAS CITY

The Woodlands Racetrack in Kansas City, Kansas (where horses race seasonally and greyhound racing is a year-round sport), promises "fun that you can bet on." Customers in the comfort of an enclosed facility watch and wager on live races as well as horse and greyhound races simulcast from tracks around the country. Greyhounds typically begin training for the sport at one year of age, but history indicates that they are born naturals; records of this racing breed date back to biblical times. "These are wonderful dogs, and they love to run," said Connie Loebsack, marketing director. Greyhounds reach speeds of up to forty miles per hour. Highly valued by their owners, each dog is allowed to compete only once every three to four days. They receive ongoing special care, such as this cool-down shower from Darren Flahive, trainer. Greyhounds retire after about five years, when they are adopted out to caring homes.

Blue Springs: Growing Community, Great Potential

Less than a half hour from downtown Kansas City, there's a newly energized, fast-growing place that knows the meaning of community.

Blue Springs was founded by pioneers in 1827 along the cooling spring waters of the Little Blue River. Growing into a friendly small town of a few thousand, Blue Springs saw its resident numbers swell from six thousand in 1970 to more than fifty-three thousand in a little over two decades.

Now, city officials, developers and planners, and community members are working together in all quadrants of the city to manage yet another growth cycle. Through their concerted efforts, Blue Springs is capitalizing on a recent annexation of two thousand acres, implementing substantial changes like the investment of $50 million in public infrastructure for sewer and water and the widening of Highway 7, which is making possible the construction of thousands of homes.

This spirit of cooperation is, to many, a complete turnaround for a city once known as too difficult to work with. Today, Blue Springs is becoming a community known for its vibrancy and its ability to initiate positive change. And with each part of the development activity, the community is resolved to maintain the balance between environment and expansion, retail and residential, progress and population.

▲ *Recreation opportunities abound throughout Blue Springs, which has 718 acres of city parks and is surrounded by thousands of acres of county and state parks. The area's numerous, large public-access lakes offer boating, fishing, and nature walks.*

CHAPTER TWO: WORKING IN GREATER KANSAS CITY

For instance, the South Development Area encompasses over fifteen hundred acres of development, comprising $1.5 billion in new capital investment that is planned to be a blend of residential, retail, and commercial properties complemented by parks and walking trails, and all of it within twenty-five miles of major employment centers.

Blue Springs is already home to some twelve hundred businesses, ranging from small mom-and-pop shops like the Soda Fountain, to light industrial firms and large global corporations. Names like Fike Corporation, Stone Container, Meyer Laboratory, Durvet, Kohls Distribution Center, and Haldex have laid the groundwork for a strong business climate that continues to welcome expanding and new companies of all sizes.

In fact, the welcome mat for new and expanding businesses provides resources to assist in new capital investment and job creation, including incentive programs such as investment tax credits and property tax abatements, job training, utility rate reductions for qualifying users, and a host of other programs.

A vast workforce in and around Blue Springs is another good reason to consider this exciting city. Some six hundred thousand employable people and tens of thousands considered underemployed—people equipped with education and skills who might consider work more befitting their experience and financial need—are located in the Blue Springs laborshed, and all within an easy commuting distance.

Of course, it isn't all about business in Blue Springs. With a median household income of more than sixty-three thousand dollars, Blue Springs is a place where businesses and people both thrive.

The community is popular for its peaceful neighborhoods, award-winning school district, and demographic mix of enterprising individuals, middle-class families, and active seniors reaching retirement years.

Affordable, diverse housing makes Blue Springs a place where individuals and families can realize their dreams of homeownership. Its quality-of-life amenities include eighteen public parks encompassing over 718 acres, featuring everything from walking paths and picnic areas to swimming pools and skating plazas. And when it's cold outside, the fun can continue indoors with a host of shopping options and local sports facilities equipped for league play or individual participation. The community also hosts year-round events and festivals that draw tens of thousands of attendees.

(continued on page 192)

BLUE SPRINGS IS A PLACE WHERE BUSINESSES AND PEOPLE BOTH THRIVE.

◀ *Among the small-town charms of Blue Springs is the Soda Fountain, a nostalgic, fully working fountain-style shop that serves ice cream, sodas, and sandwiches. This fun and tasty establishment is nestled in the heart of downtown Blue Springs, which is now enjoying a rebirth of revitalization.*

CITY OF BLUE SPRINGS

PHOTO BY DENNIS KEIM

(continued from page 191)

As the seventh-largest suburb in metropolitan Kansas City, Blue Springs is close to all the excitement of the big city, from metropolitan-style shopping and nightlife to world-class sporting events to international conventions. But when the day is through, the people of Blue Springs agree: there's no other place they'd rather call home. ♦

▲ *The Blue Springs BMX track, a unique asset of the Parks and Recreation Department, is one of the few programs of its kind nationally. Every year the track is used by local, regional, and national riders competing in tournaments.*

PHOTO BY ERIC FRANCIS

◄ *The Blue Springs Police Department oversees twenty-one square miles, using traditional vehicle patrols as well as bike patrol units to bring officers in closer contact with activities. Citizen surveys consistently rank safety as one of the best services that the city provides.*

CHAPTER TWO: WORKING IN GREATER KANSAS CITY

PHOTO BY ALAN S. WEINER

Blue Springs residents enjoy a very active calendar of arts programs throughout the community. One volunteer-led group, the Blue Springs City Theatre, undertakes several productions each year. The group invites members of the community to participate, either auditioning for roles onstage or lending a hand behind the scenes.

Top Innovations, Inc.: Products to Streamline Your Life

Many U.S. television viewers are familiar with the Singer Handy Stitch. However, they probably don't know that this intriguing household product was introduced to the American public by Benny Lee, owner of a Taiwan-based trading company. In 1987, Lee founded his own company—Top Innovations Inc.—and supplied several successful products, including the famous Ginsu 2000 on TV in the mid-1990s.

"Ours is a company with international presence that develops and sells products that make consumers' households work easier, faster, and more effective, *and* that are helpful to the environment," said Benny Lee. The company's extensive product line includes a variety of steam cleaners, hand-held vacuums, and garment steamers manufactured under the McCulloch and Steam-Fast brand names. Studies show that these cleaning products destroy a higher percentage of germs than cleaners such as Lysol, with the added benefit of not chemically compromising the environment. An upcoming line of "steam vacuums" sounds like a dream to allergy sufferers: as the steam kills bacteria and dust mites, it also vacuums up their remains, which act as allergens if left behind. A health-and-fitness line offers great promise for people who are "too busy to exercise." Fitness Stride, which uses a series of Velcro

▲ *Benny Lee, CEO of Top Innovations, pictured in his forty-seven-thousand-square-foot, newly expanded warehouse. The company ships consumer products daily throughout the United States.*

straps and power bands, allows the wearer to work out specific muscle groups while going about everyday chores, doing yard work, or even working at the office.

Top Innovations products are a mainstay on QVC and Amazon, and are available online with retailers such as Target and Wal-Mart. Many of these products have been awarded the coveted Good Housekeeping Seal, which can be earned only after meeting rigorous testing standards that mean much to consumers in terms of service and extended warranties.

The company and its staff recently celebrated the dedication of a newly renovated forty-seven-thousand-square-foot warehouse and office headquarters in Kansas City, Missouri. Lee also spends much of his time in the Far East, diligently developing new products while visiting offices in Hong Kong, Taipei, and Barcelona, and factories in China. As the CEO of a growing global enterprise, Lee is equally zealous about contributing to the local community. He and his wife, Edith, won the 2005 Northland Community Foundation Pinnacle Award for their promotion of classical music at Park University, and the couple contributes regularly to many local charities. Benny Lee also serves on the board of the Heart of America United Way and is a board member of the Greater Kansas City Chamber of Commerce. He is a trustee of Park University, and is a board member of the International Relations Council.

He and the employees at Top Innovations strive to achieve a balance so that they continually attain their ultimate accomplishment: "a totally satisfied end-user, customer, and employee." ♦

> OUR ULTIMATE ACCOMPLISHMENT IS "A TOTALLY SATISFIED END-USER, CUSTOMER, AND EMPLOYEE."

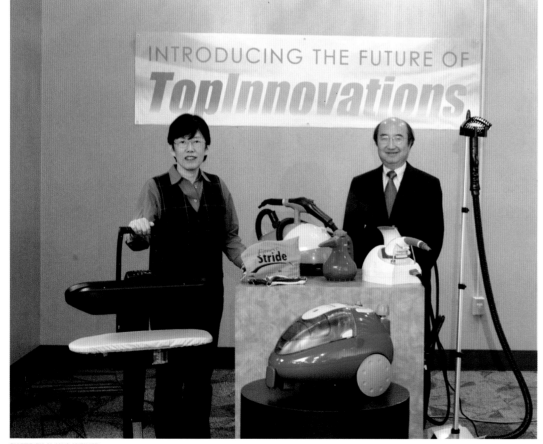
PHOTO BY ALAN S. WEINER

◀ Owners Edith and Benny Lee display a few of their consumer products, including SteamFast and McCulloch steam products, and Fitness Stride Health & Wellness items.

PHOTO BY MARIO MORGADO

PHOTO BY MARIO MORGADO

196 | GREATER KANSAS CITY: UNLIMITED POSSIBILITIES

CHAPTER TWO: WORKING IN GREATER KANSAS CITY

Started in 1827 as a frontier outpost playing a key role in the nation's westward expansion, Fort Leavenworth remains the oldest continuously operating military installation west of the Mississippi River. While its initial missions were to protect caravans and help to relocate American Indians, today the base is often referred to as the "Intellectual Center of the Army," playing a major role in the changing operations of this arm of service. As home to the U.S. Army's Combined Arms Center, the base directs army doctrine revisions, oversees officer instruction at numerous locations nationwide, provides support for large training exercises, collects and disseminates training and contingency operations information, and guides the integration of new systems and structural changes.

◀ *Founded in 1889, Shook, Hardy & Bacon is an international law firm with a Kansas City office located at Crown Center, and eight other offices strategically located throughout the world.*

Shook, Hardy & Bacon LLP: The Firm with a History of Excellence

You might expect a law firm established in 1889 to have a long and impressive history. However, in talking with chairman John Murphy, it is quickly apparent that Shook, Hardy & Bacon LLP exceeds expectations. History is just one example. Missouri's governor appointed one of the firm's first partners as a special prosecutor in the trial of Frank James. In the 1930s, another partner, Edgar Shook, led the battle against the Pendergast political machine in Kansas City.

This zeal for civic improvement and community involvement is still at the heart of SHB. Today the firm has its largest office in Kansas City, locations in six other major cities in the United States, as well as offices in Geneva, Switzerland, and London, England.

"Being a full-service, international firm allows us to give our clients the

CHAPTER TWO: WORKING IN GREATER KANSAS CITY

highest level of service. Many of them have global interests, and we represent them in these international markets," said Murphy.

The International Who's Who of Business Lawyers has recognized SHB as the Global Product Liability Law Firm of the Year. As their vision states, the firm is "committed to being the best in the world at providing creative and practical solutions at unsurpassed value."

An impressive number of lawyers in the Kansas City office appear in *The Best Lawyers in America*. "This is significant," said Murphy, "because this honor is bestowed by their peers. It means you are a lawyer's lawyer."

However, the smartest lawyer cannot function without a top-notch staff. "Often the first contact with our firm is with the switchboard or the receptionist, so every person is important," Murphy continued. "One of our strongest assets is the staff and the analysts and researchers who support our attorneys."

The staff is another example of how SHB goes beyond expectations. In 2006, SHB received the Law Firm Diversity Award from the Defense Research Institute. The firm's commitment to diversity is demonstrated through politics, practices, and civic contributions. And it is in the area of civic contributions that SHB really shines.

Recently the firm won a landmark civil rights victory for foster children. The fees awarded were donated to Midwest Foster Care and Adoption Assistance to fund a new program called "Lawyers for Kids," which will help disabled and foster children and their families in the Kansas City area. SHB also has one of the nation's top one hundred pro bono programs. One hundred fifty firm attorneys serve on nonprofit boards, and in 2006 SHB donated a total of more than five hundred thousand dollars to two hundred different organizations. Local SHB volunteers contribute their time and resources to a long list of worthy causes.

Exceeding expectations is SHB's hallmark. The final words of their vision refer to a work environment "where everyone is respected, feels appreciated, and experiences fulfillment and enjoyment through meaningful personal contributions." ♦

> "SHOOK, HARDY & BACON IS COMMITTED TO BEING THE BEST IN THE WORLD AT PROVIDING CREATIVE AND PRACTICAL SOLUTIONS AT UNSURPASSED VALUE."

◀ *Rob Adams, George Wolf, Madeleine McDonough, and John Murphy (standing, left to right), and Marty Warren, Ken Reilly, and Bruce Tepikian (seated, left to right) are among the members of the firm's Executive Committee, which sets the strategic course for Shook, Hardy & Bacon.*

In 1997, realizing that he could no longer find old-timey marbles for inclusion with his handmade game boards, Bruce Breslow (right) opened the Moon Marble Company and started making them himself. Here, he demonstrates the art and science of handmade glass marble making while two young visitors decide which ones they want for their own collections. In addition to handmade marbles by Breslow and other artisans, Moon Marble sells machine-made and bulk marbles, wooden game boards, vintage tin windup toys, glass ornaments, and a variety of novelty, prank, and magic items. Bruce is available to demonstrate his skills as well. With names like Alien Swirl, Blue Spotty, Tie Die Snippy, and Dark Side of the Moon, his one-of-a-kind creations are delightful reminders of why marbles remain a perennial favorite with kids of all ages.

CHAPTER TWO: WORKING IN GREATER KANSAS CITY

Internationally known miniature artist William R. Robertson has been professionally crafting one-twelfth-sized reproductions of real objects for the last thirty years, an interest that goes back to childhood. From pure gold and other metals to exotic woods, Robertson makes exact replicas of period items using the techniques and tools of the day. "I use the techniques of a whole lot of different trades in any given piece," he says. "I study the way things were done in the workshop two or three hundred years ago and then duplicate it." Collectors and museums from across the globe turn to Robertson to create the minuscule objects, which can take from a few hours to literally years to make. Here Robertson holds a fully functional miniature of a Louis XV microscope made by Claude-Simeon Passemant, circa 1760. Crafted in 24K gold and an eighteenth-century type of sharkskin known as shagreen, it is made up of approximately 125 parts. Also shown below, from left: a gentleman's tool chest, circa 1770; a French ladies tabletop spinning wheel, circa 1790; three brass candlesticks; a nineteenth-century Dolland refracting telescope; a watercolor paint set with real paints, brushes, and an ivory pallet; an eighteenth-century surveyor's compass; and an eighteenth-century English brass ink stand.

Key Companies & Associates LLC: From Idea to Empire in the Land of Entrepreneurial Opportunity

The Greater Kansas City area's reputation as an entrepreneurs' paradise is a well-deserved one. Just ask Joe and Judy Roetheli, who—after exhausting all existing answers for their beloved dog's bad breath—took it upon themselves to create their own solution. The result was Greenies® treats, a blend of proteins, vitamins, and minerals in a dog chew designed to clean teeth, freshen breath, and help maintain diet.

That was 1996, and only ten years later the Roethelis sold their multimillion-dollar venture to pet food giant Mars Inc. At that time, S&M NuTec had rocketed to become the eighth-largest pet food/treat company in the world, having sold more than 750 million Greenies® chews in sixty countries—enough to circle the globe 1.6 times when placed end to end.

So how does a young company, whose leaders worked in public service and had no business experience, achieve such a record of success? The Roethelis will tell you it's because Kansas City is a place where anyone can transform a dream into reality.

For starters, they credit a team that

▲ Max, Joe and Judy Roetheli's dog, seems to study the company's awards while salivating over Greenies®, the teeth-cleaning and breath-freshening product that launched what has become the eighth-largest pet food company in the world.

▶ Judy and Joe Roetheli, entrepreneurs whose business venture demonstrates that the American dream is still alive and available to creative thinkers who are willing to take risks.

showed them unequaled loyalty when presented with a working environment that encouraged input and rewarded creativity. They also found the same midwestern work ethic in their local manufacturing, packaging, warehousing, and shipping outsourcers and considered those hundreds of local employees to be an extension of their own operations.

Another part of the genius was a creative product that dogs loved, which performed a function and met a true market need.

Of course, a centralized location with an abundance of raw materials and a superior transportation system gave their operations a boost. Moreover, the Roethelis readily recognize the fact that strong business partnerships forged early on with local suppliers and molders were a significant factor in their success.

In addition, they found mentoring to be a valuable Kansas City asset, from organizations like the Kauffman Foundation, the Henry W. Bloch School of Business and Public Administration at the University of Missouri–Kansas City, the Helzberg Entrepreneurial Mentoring Program, and the Greater Kansas City Chamber of Commerce.

It also helps that the Roethelis racked up a few awards along the way, including: Euromonitor's ranking of Greenies® as the top-selling brand of dog treats in the United States and the second-best-seller in the world; a highly coveted Mr. K. Award from the Greater Kansas City Chamber of Commerce for being the best small business; a couple of American Business Association's Stevie Awards for Best Overall Company with Under 100 Employees and Best Packaging Design in Marketing; an Exporter of the Year Award from the National District Export Council; and an Entrepreneur of the Year in Manufacturing for Western Missouri and Kansas given by Ernst & Young.

Today, the Roethelis have launched several new endeavors: Key Companies & Associates, Simple Man Products™, MySYZYGY, and the Roetheli Lil' Red Foundation. For any entrepreneur considering their options, the Roethelis have a simple message: dream, work hard, have fun, and think Kansas City. ♦

> THE ROETHELIS HAVE A SIMPLE MESSAGE: DREAM, WORK HARD, HAVE FUN, AND THINK KANSAS CITY.

PHOTO BY DENNIS KEIM

Since John McDonald started his Boulevard Brewing Company in 1989, he and his crew have focused on one thing: brewing great specialty beer. That single-mindedness has made Boulevard Brewing Company the largest brewer of its kind in the Midwest. Brewing four staple beers year-round, and other labels for each season, Boulevard's beers have become regional favorites of beer drinkers who like full flavor with their bottle of brew. On Fridays and Saturdays, by reservation, visitors can tour the brewing facility at 2501 Southwest Boulevard to hear the company's history, see how beer is made, and take a taste of the brewer's various beers.

CHAPTER TWO: WORKING IN GREATER KANSAS CITY

"Hi, may I help you?" is the trademark of Gates and Sons Bar-B-Q and the customer greeting that every employee learns at the Gates College of Bar-B-Que Knowledge, also known as Rib Tech. "Everybody who works in the restaurant must complete specific training courses at Rib Tech," said Arzelia Gates, community relations. Gates's grandfather George W. Gates perfected the barbecue sauce that made the original restaurant famous, and her father Ollie W. Gates defined the measurements for the secret recipe. "It's spicy, not hot, and very flavorful," said Gates. Gates's sauces, seasonings, and barbecue are so tasty that the first restaurant, which opened in 1946 in Kansas City, Missouri, has expanded to six locations.

PHOTO BY DENNIS KEIM

Forget those fancy Food Network cook-offs with their exotic ingredients and esoteric techniques. Barbecue challenges, America's bastion of culinary egalitarianism, require only a sauce, a rub, and a penchant for fire. And lots of enthusiasm, like that shown by participants Nick Kure (left) and Mark Shniderson, who enjoy the competitive spirit and party atmosphere. Most barbecue challenges are organized by the Kansas City Barbeque Society, the world's largest barbecue promotional organization, which each year sanctions over 250 competitions nationwide. Close to 30 are located within a one-hundred-mile radius of Kansas City alone.

PHOTO BY SCOTT INDERMAUR

At the crossroad of three interstates—I-29, I-35, and I-70—Kansas City has more highway miles per capita than any other city. In fact, Kansas City exists because of the waterway that separates its two states, used as its earliest source of commerce. When a railroad bridge was built across the river in the 1860s, it spurred the city's growth and put it forever on the nation's map. Today, the city remains the nation's second-largest rail center, and transportation and logistics industries continue to be big players locally.

CHAPTER TWO: WORKING IN GREATER KANSAS CITY

Once the corporate headquarters of Trans World Airlines, the building at 1735 Baltimore Avenue, in Kansas City's Crossroads Arts District, now houses the renowned Barkley advertising agency. When the airline moved to New York City in 1958, the building was converted to a training facility for flight attendants, a purpose it served for a little more than a decade. For a number of years, the building's future hung in the balance, and in fact, it had been vacant for awhile when, in 2002, it was listed on the National Register of Historic Places. Today, Barkley is returning the building's glory, making use of its theater, art gallery, and rooftop observation deck. The agency has also replicated and reinstalled the TWA Moonliner on the building's roof. The original forty-foot rocket, designed by the Walt Disney creative team, stood on the roof as a symbol of TWA owner Howard Hughes's dream of taking the airline into the realm of space travel.

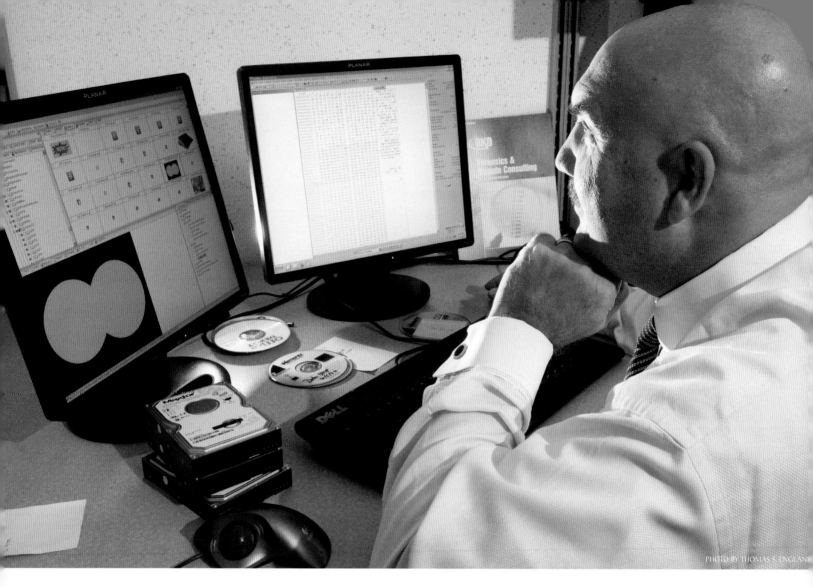

BKD Stands Apart with Unmatched Client Service

BKD, LLP, is one of the largest CPA and advisory firms in Kansas City, and one of the ten largest firms in the country, presently located in twenty-seven offices across eleven states. As the CPA and advisory firm of choice for growing companies and high-net-worth individuals, BKD certainly has the resources to support the business needs and concerns of its clients.

What truly distinguishes BKD from other CPA firms is the way it treats clients. "When providing a service, an intangible, it can be difficult for the recipient of that service to distinguish yours from someone else's," said Doug Gaston, partner-in-charge. "That's why we place such an emphasis on client service."

In fact, when it comes to client service, BKD literally wrote the book. Its publication, *The BKD Experience: Unmatched Client Service*, provides a roadmap of sorts for everyone at the firm.

Spelled out within the book's pages is the definition of PRIDE, an acronym for the values of passion, respect, integrity, discipline, and excellence. More than defining expectations, these words serve as cues to the basics. "We expect our professionals to return phone calls promptly, be on time for meetings, and be courteous," said Gaston. "None of these concepts are complex, but our emphasis encourages us never to lose sight of our core values."

▲ *John Mallery, managing consultant for BKD Forensics & Dispute Consulting (FDC), in the FDC lab where he retrieves data from hard drives, databases, and networks to locate the smoking gun in an investigation.*

CHAPTER TWO: WORKING IN GREATER KANSAS CITY

This strong focus on the basic principles of good business was affirmed when BKD received the results of a recent comparative study by an independent research firm. Based on the results, BKD clients report they are significantly more satisfied with their service than clients of other top U.S. accounting firms. Furthermore, BKD's overall results are higher than those of any other CPA firm that the researcher has surveyed in the past twenty-five years.

In Kansas City, BKD, LLP, delivers innovative, timely, and affordable solutions to thousands of clients in a wide spectrum of industries. The firm's highly specialized professionals dedicate themselves not only to specific industries, but also to strategic areas within those industries. In addition, the Kansas City office houses several of the firm's largest niche groups and specialized consultants who bring added value by assisting businesses in operating more efficiently and profitably.

Through its WealthPlan division, BKD addresses the needs of individual clients, including the chief executives of its diverse client base. "We recognize if we don't take care of their personal needs, we're not providing a comprehensive service," said Gaston.

BKD made a conscious decision in the late 1990s to keep its offices in the city's center. "There were several locations that we could have chosen, but as one of the larger firms in the city, we felt an obligation to support downtown," said Gaston. "We're glad we made the decision to stay. With everything that is happening in downtown Kansas City, it's going to be an exciting place and we'll be right in the center of it."

Through its own foundation, BKD also distributes dollars back to the Kansas City community to support a range of charitable organizations. "We encourage our people to be involved in at least one charitable or civic organization," said Gaston. "It's a way of giving back to the community. You interact

(continued on page 210)

"OUR EMPHASIS ENCOURAGES US NEVER TO LOSE SIGHT OF OUR CORE VALUES."

PHOTO BY THOMAS S. ENGLAND

◀ *One of BKD's many contributions to the community: a sponsorship for Treads & Threads, a benefit for the University of Kansas Hospital. (Left to right) Sue Brammer, partner, and husband Steve; Dan Beattie, partner; Anita Hempy; and Ted Hempy, partner.*

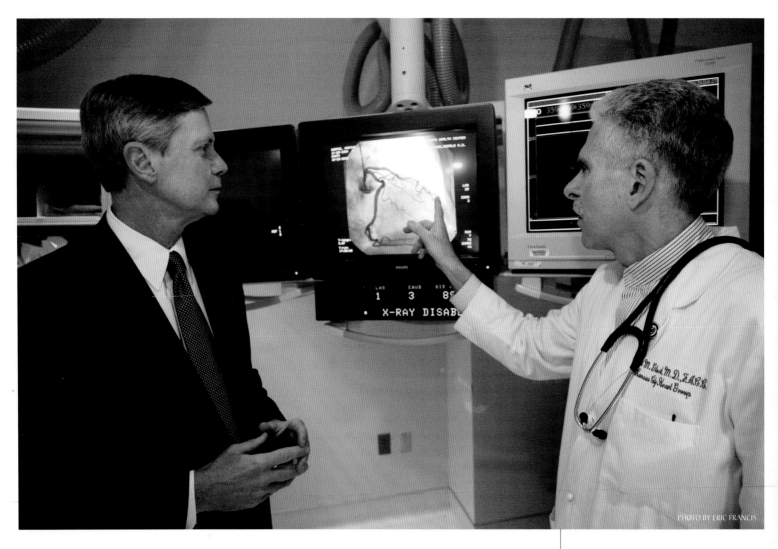

(continued from page 209)

with top people and broaden your view of the world, which allows you to be exposed to key developments affecting our city."

For BKD, the bottom line is that being a part of Kansas City has always made good sense. ♦

▲ *Partner-in-charge Doug Gaston listens to Dr. Robert Glueck of Kansas City Heart Group explain technology used in assessing the health of patients' hearts.*

◄ *A valuable BKD client, Murphy Hoffman Company displays an impressive line of Kenworth trucks at its Kansas City dealership.*

CHAPTER TWO: WORKING IN GREATER KANSAS CITY

PHOTO BY ERIC FRANCIS

The Central Jackson County Fire Protection District is consistently rated a top value by Blue Springs residents, who note the professionalism demonstrated in the delivery of fire and emergency medical services throughout most of Blue Springs. More than one hundred full-time professionals make up the group, staffing four engine companies, one ladder company, and three life-support ambulances.

PHOTO BY SCOTT INDERMAUR

Over one hundred Burns & McDonnell employee-owners helped renovate a home for TLC for Children and Families. TLC provides housing and social services for boys and girls ages eleven to eighteen who have experienced abuse, neglect, or delinquency. Pictured here are just a few of the employee-owners who donated their time over a weekend.

Kansas City Corporate Challenge is an athletic competition that allows companies to compete in a variety of sporting events. One of the largest such events in the country, more than twenty thousand individuals participate. The athletes from Burns & McDonnell are giving their all in the tug-of-war. The engineering firm has won its division each of the last three years.

CHAPTER TWO: WORKING IN GREATER KANSAS CITY

Located at 1830 Main Street in downtown Kansas City, Michael's Fine Clothes for Men is one of those retail rarities: a small family business that's made it big. In fact, some of its customers have been coming to Michael's not for years, but for decades. One reason is the store's large selection of fine men's clothing and accessories. Another is the tradition of personalized service that goes back to the shop's founding in 1905 by Russian immigrant Michael Novorr. Today grandson Keith Novorr runs the store and has called his work not a job, but a legacy.

PHOTO BY MARIO MORGADO

PHOTO BY THOMAS S. ENGLAND

It's the beginning of August and time to . . . put up Christmas lights? Absolutely, and not a minute too soon when you have to hang 287,000 of them. Mike Burks of Trimble, Missouri, rode a sixty-six-foot boom lift to get to the top of the Clock Tower on Triangle Block at Country Club Plaza. He and a crew of five other electricians will work diligently to get the lights in place to be turned on at Thanksgiving. When the work is done, all the towers and the rooflines of all the buildings on the Plaza will be illuminated. Mike has been involved in this project for the past ten years. The lights stay up until the second Sunday in January, when—guess who—starts taking them down. Mike and an assistant work until the first of May to remove the lights, and almost as soon as that's finished, it's time to start over.

GREATER KANSAS CITY: UNLIMITED POSSIBILITIES | 213

PHOTO BY MARIO MORGADO

Progress is in the air over Kansas City. Since 2000, its downtown has undergone an estimated $4 billion in renovations, making it one of the most desirable places in the region to live and work. Diverse as well as booming, it is home to historic buildings and new constructions of glass and steel, corporate headquarters and specialty retail establishments, art centers, and residential enclaves.

CHAPTER TWO: WORKING IN GREATER KANSAS CITY

PHOTO BY MARIO MORGADO

Enjoying
KC Greater Kansas City

Great music can't be made in a vacuum. It needs to be nurtured, encouraged, and provided with just the right atmosphere in which to flourish. Almost since the city's inception, Kansas City has provided the people and places that help make great music happen, from piano bars and musical theaters to savvy promoters and dedicated fans.

Looking for something exciting to do? Greater Kansas City has it.

A city of four distinct seasons, Kansas City sports more than two hundred parks, featuring everything from baseball diamonds and tennis courts to swimming pools and art exhibits. Whether you want to walk, run, fish, or play a round of golf, there's a place to do it in and around Kansas City.

Looking for a little spectator action? Residents on both sides of the state line lay claim to Royals baseball, Chiefs football, and NASCAR excitement at the Kansas Speedway. The American Royal Complex draws acts from far and wide for concerts and events and, with its Hale and Sprint Center arenas, sets the stage for horse shows and rodeos for riders of all ages.

There's a rich history just waiting to be explored at places like the Negro Leagues Baseball Museum, the museums at 18th & Vine, or the Strawberry Hill Museum. Even exploration of the future is in store at Science City at Union Station.

When it comes to shopping, chances are, if you want it, Kansas City has it. From strip malls to big-box stores to the Legends at Village West, the selections are endless. Head over to Crown Center for three levels of shops and a tour of the Hallmark Visitors Center to see the history of this world leader in the card industry. A trip to the Country Club Plaza is always a special treat, but especially during the holidays when a spectacular lighting celebration illuminates the night.

If a night on the town is calling, there's enough to keep you busy until dawn. Start with a meal at one of the area's multiple restaurant options, featuring choices ranging from Kansas City beef to a world of ethnic cuisine. Then delight in a performance by the world-class Kansas City Symphony or guffaw at the antics of the stage acts over in the Blue Springs City Theatre. You can get a meal and a show, featuring some well-known performers, at the New Theatre Restaurant. Or maybe just try a roll of the dice in one of the local riverboat casinos.

Thousands make Kansas City a favorite getaway, taking advantage of the more than 120 hotels and twenty-six thousand rooms available for a night or a weekend stay. They know that day and night, Kansas City is a place where something is always going on.

AMC ENTERTAINMENT INC.	220
APPLEBEE'S INTERNATIONAL, INC.	254
CHASE SUITES HOTELS	258
COURTYARD BY MARRIOTT COUNTRY CLUB PLAZA	224
EXECUTIVE AIRSHARE	230
GARMIN INTERNATIONAL, INC.	288
HALLMARK CARDS, INC.	244
THE HNTB COMPANIES	278
HOEFER WYSOCKI ARCHITECTS, LLC	314
HOTEL PHILLIPS	270
KANSAS CITY MARRIOTT DOWNTOWN	248
KANSAS CITY SOUTHERN	298
KANSAS CITY SYMPHONY	264
KESSINGER/HUNTER & COMPANY	294
SPRINT	236
WDS MARKETING & PUBLIC RELATIONS	282

No matter your level of fan dedication, a day spent at a Kansas City Royals baseball game is a guaranteed good time. A Central Division American League team, the Kansas City Royals are a "True Blue Tradition," part of Kansas City's professional baseball legacy that goes back to 1883. When the Kansas City Athletics moved to Oakland after the 1967 season, a new team was awarded to the city to start play in 1969. In 2007, the Royals played their thirty-fifth season in the venerable Kaufman Stadium. Kaufman will soon undergo a major renovation, expanding its seating and adding a host of new amenities, including a brand-new entertainment plaza.

AMC Entertainment Inc. Pioneers the Moviegoing Industry

Each year, billions of people choose to leave their homes and go out to the movies. There's just something magical about the experience, and for nearly a century, AMC Entertainment Inc. has been there to make that magic happen.

"Moviegoing is the most popular form of out-of-home entertainment," says AMC®'s chairman, CEO, and president, Peter C. Brown. "The sound, the visual components, the quality, comfort, and big-screen experience can't be beat. You simply can't replicate it anywhere else."

Founded in 1920 in Kansas City by Edward Durwood, AMC is one of the oldest motion picture exhibition companies in the industry. Defined by a rich history of innovation and commitment to guest service, it is also one of the country's largest exhibitors. AMC has interests in more than four hundred theatres in thirty states and the District of Columbia, and ten countries including the United States. AMC's theatres attract approximately 240 million guests a year.

A pioneer credited with many of the industry's "firsts," AMC introduced the multiplex concept in the United States in 1963, opening the world's first shopping center "twin" theatre at Ward Parkway Mall in Kansas City. Three years later, AMC introduced the first four-screen theatre, followed by the first six-screen theatre in 1969. The multiple-screen concept reached its height in 1995 when the company opened the country's first megaplex, the

▲ *AMC's mission is to provide its guests with the best possible out-of-home entertainment experience. The company and its associates passionately believe in ensuring friendly, helpful, and fast service for all guests and are known for their deep commitment to guest satisfaction, innovation, quality of workplace, and performance.*

CHAPTER THREE: ENJOYING GREATER KANSAS CITY

AMC Grand 24 in Dallas, Texas. With a mission to provide its guests with the best possible out-of-home entertainment experience, AMC also created innovations such as the cupholder armrest, LoveSeat®-style seats, online ticketing services, and the industry's first stored-value gift card.

AMC focuses on what it calls the "Three P's": Place, Programming, and Price. "We know great locations and quality assets are key to attracting guests," says Brown. "We build high-quality, state-of-the-art theatres in prime locations only." Programming is geared to include first-run movies, the latest in hard-to-find specialty films, and alternative content. AMC continuously looks for ways to make the experience of going to movies at its theatres more accessible to more people.

"We have a saying," Brown continues. "'Take care of the guest and you won't have to worry about the rest.' Guest satisfaction is built into our company DNA, and we work hard every day to ensure that our guests have an unparalleled experience when they visit us."

AMC makes going to the movies fun and beneficial by directing a portion of its resources each year to numerous local and national health-care, educational, and community enrichment projects. It is also one of downtown Kansas City's greatest champions. With company headquarters located downtown near where Durwood founded his first theatre, AMC is dedicated to revitalization efforts in the area. In July 2005, AMC and the Cordish Company entered into a joint venture to refurbish two of AMC's historic downtown venues. Today, the Midland and Mainstreet theatres anchor the northwest and southwest corners of the city's Power & Light District.

Says Brown, "We intend to be a prominent leader in defining the next generation of the movie theatre business; that is a continuum of our history." ♦

"TAKE CARE OF THE GUEST AND YOU WON'T HAVE TO WORRY ABOUT THE REST."

◀ *AMC's worldwide headquarters, located in the heart of downtown Kansas City, Missouri. AMC was founded in the early 1920s and has maintained its headquarters in Kansas City throughout the company's more than eighty-five-year history. Among other historical company innovations, the nation's first multiplex was opened at what is today's AMC Ward Parkway 14 theatre.*

Since 1957, the Kansas City Ballet has danced its way into the hearts of enthusiasts from around the region. From classical favorites to modern works, the Kansas City Ballet's performances continue to delight audiences of all ages. Whether they are on stage, conducting a workshop, or sharing their talents with youth in a classroom, the Kansas City Ballet's dancers know how to make ballet an exhilarating experience.

CHAPTER THREE: ENJOYING GREATER KANSAS CITY

Courtyard by Marriott in the Heart of History

Since it was originally erected in 1925, the six-floor building housing the Courtyard by Marriott Country Club Plaza has been a premier destination in Kansas City. It was originally the Park Lane Apartments, one of the city's distinctive southside residential hotels. The Winston Hotel Group purchased the building, which spared it from destruction, and after extensive renovations, it was opened as a Courtyard by Marriott in April 2006. The hotel is on the National Register of Historic Places because of its location, design, setting, materials, and workmanship.

Today's guests enjoy the spectacular views of the famed Country Club Plaza, the J. C. Nichols Fountain, and Mill Creek Park. The building's Mission Style design reflects the overall architecture of the Country Club Plaza. Visitors are right in the center of the trendy, sophisticated Kansas City dining, shopping, and cultural scene.

The Plaza itself is a virtual museum of romantic Spanish architecture and European art. It was designed in 1922 as the nation's first suburban shopping district. Guests staying at the Courtyard by Marriott are just steps away from restaurants, bars and nightclubs, and over one hundred retail stores. The Nelson-Atkins Museum of Art and the Westport area are within a one-mile radius of the hotel. Downtown Kansas

▲ *Courtyards by Marriott are known as the hotels designed by business travelers. However, they are equally dedicated to making pleasure travelers feel at home. From the Marriott's luxurious new bedding package to a healthy breakfast buffet or a glass of wine before dinner, the Marriott has thought of everything.*

CHAPTER THREE: ENJOYING GREATER KANSAS CITY

City is four miles north, with the Truman Sports Complex ten miles northeast.

Each of the Courtyard by Marriott's 123 rooms has a charm of its own—equipped with plasma televisions with free HBO premium channels and pay movies, a comfortable sitting area, refrigerators, microwaves, and in-room coffee and tea service... all this plus Marriott's luxurious new bedding package.

Business travelers, of course, enjoy all those amenities, and they will especially appreciate the fact that this hotel was designed by business travelers who know what it takes to make business a pleasure. The Courtyard by Marriott Country Club Plaza has 924 square feet of meeting space, comprising a 644-square-foot ballroom and a 280-square-foot boardroom. The hotel is convenient to all major businesses and only minutes from the Kansas City International Airport. For travelers who need to get some serious work done, each room includes high-speed Internet access; two dataport telephones with speaker and voicemail; a large, well-lighted desk; and an ergonomic chair for added comfort.

The Courtyard Café offers breakfast daily, and the lobby bar and lounge are a great place to meet, greet, and network with business contacts, friends, or family. Need to relax after a day of meetings or shopping? The fully equipped fitness center, the heated outdoor swimming pool, and the indoor whirlpool provide even more recreational options.

Of course, the Courtyard by Marriott Country Club Plaza is perfect for vacationers and business travelers, but it also offers packages designed for special occasions. Enjoy a romantic evening with deluxe accommodations, roses, champagne, dinner for two at the M&S Grill, and even a carriage ride through Country Club Plaza. Or perhaps its time for the ladies to treat themselves and their friends to "It's a Girl

(continued on page 226)

> VISITORS ARE RIGHT IN THE CENTER OF THE TRENDY, SOPHISTICATED KANSAS CITY DINING, SHOPPING, AND CULTURAL SCENE.

PHOTO BY SCOTT INDERMAUR

◀ *The Courtyard Kansas City is located in the heart of the famed Country Club Plaza across from Mill Creek Park and the beautiful J. C. Nichols Fountain. It is the perfect location for dining, shopping, exploring the history of the city, or taking in the arts scene.*

COURTYARD BY MARRIOTT COUNTRY CLUB PLAZA

The guest rooms and suites at the Courtyard by Marriott Country Club Plaza blend business-friendly features like complimentary high-speed Internet access, plasma televisions, microwaves, and refrigerators. Well-lighted desks are convenient, and the couches are oh-so-comfortable.

PHOTO BY SCOTT INDERMAUR

(continued from page 225)

Thing." Shopping on the Plaza is a mini-vacation guaranteed to please.

Throughout its history, from its early days as the Park Lane Apartments to the present, the Courtyard by Marriott Country Club Plaza has played a vital part in Kansas City culture. The hotel continues to be involved in the community through its support of the Convention and Visitors Association, the Chamber of Commerce, the Hotel and Lodging Association, and the Kansas City Meeting Planner. ♦

The Courtyard Country Club Plaza takes the hassle out of business get-togethers. The Courtyard's meeting professionals approach every event, large or small, with the expertise and dedication to make your guests comfortable.

PHOTO BY SCOTT INDERMAUR

CHAPTER THREE: ENJOYING GREATER KANSAS CITY

Saddle up, round 'em up, and move 'em . . . downtown? This was the theme of civic and business leaders as they drove more than twenty-five cattle through the city streets of Kansas City, Missouri. Reviving a tradition that had been on hold for several years, city leaders and the American Royal Association (a not-for-profit organization that benefits youth and education) relived the city's Old West heritage with the American Royal Cattle Drive, sponsored by the Sprint Center. The cattle made their way down Wyandotte Avenue amidst a much-changed landscape in a drive that culminated at Barney Allis Plaza. "Ranchers from all over used to drive cattle to the Kansas City Stockyards to sell. We hold this event to celebrate Kansas City's rich agricultural background and also to kick off our livestock show, which is entering its 108th year," said Jody Holland, marketing/event coordinator for the American Royal Association.

The Kansas City Bicycle Club not only provides great riding opportunities throughout the area, but also builds camaraderie among like-minded enthusiasts. Established in 1963 to promote all aspects of cycling, the nonprofit club sponsors a variety of rides for beginners on up. Affiliated with the League of American Bicyclists, a national touring and legislative organization; the United States Cycling Federation, the nation's amateur racing organization; and Adventure Cycling, a nonprofit service group for touring bicyclists, the club also promotes the USCF-sanctioned Kansas City Bicycle Club Racing Team and sponsors the annual Tour of Kansas City race each fall.

Have Your World on Your Schedule with Executive AirShare

"Mr. Taylor is seeing a client off. He'll be here in just a minute."

When was the last time you left on a flight and had the president and CEO of the fractional aircraft company walk you to your plane, shake your hand, and wish you a pleasant trip? If you were a customer of Executive AirShare, you wouldn't find that surprising at all.

Executive AirShare is a regional, fractional aircraft company selling ownership shares in private aircraft. "Once you own a share in one of our aircraft, you tell us when and where you want to go, and we do the rest. It's as close as you can get to owning your own plane without the huge investment or the hassle," says Bob Taylor. "You fly in a plane you're familiar with, with pilots you know, and you leave from an airport near your home."

Executive AirShare's extraordinary growth is fueled in part by the increased delays, extensive security, canceled flights, lost luggage, and reduced in-flight service of commercial airlines. "Convenience and service are extremely important, but we put safety above everything else," explains Taylor. "Safety is our highest priority."

▲ *As part of the personal service customers have come to expect, Executive AirShare's CEO Bob Taylor greets two executives as they board the plane. Chief Pilot Troy VanBuskirk is there to load the baggage before they take off for their destination.*

CHAPTER THREE: ENJOYING GREATER KANSAS CITY

PHOTO BY ERIC FRANCIS

Customers also appreciate the time they save. The Beechjet 400A features leather club seating for up to seven passengers. Its cruising speed is 515 miles per hour with a range of sixteen hundred miles. Stowable worktables and a relaxed, quiet atmosphere make travel time more productive. Another important advantage is that travelers get to spend more quality time at home with their families. As an Executive AirShare customer, it truly is "Your World on Your Schedule."

"Our choice was to purchase a share in the Beechcraft Super King Air 350," says Gregory Silvers, chief operating officer of Entertainment Properties of Kansas City. "Our whole team can go, and what used to take three or four days now takes one. We can visit several cities in the same day, and if a meeting runs longer than expected, we don't have to worry about missing a flight. Our plane is there and waiting."

Owning a share in one plane gives clients the flexibility of using any plane in the Executive AirShare fleet. "Sometimes it makes more sense to use the King Air C90B, which is perfect for mid-range flights," Silvers reports.

Flights from Kansas City leave from Charles B. Wheeler Downtown Airport or Johnson County Executive Airport. Other departure cities include Tulsa, Dallas/Ft. Worth, and Wichita, Kansas. "We plan to add more service locations and to purchase several light jets and a mid-sized jet," states Taylor.

Executive AirShare is actively involved in the communities it serves. Taylor is a member of the Greater Kansas City Chamber of Commerce, a trustee of the University of Kansas Endowment Fund, and a member of the Board of Advisors of the KU School of Business.

"I enjoy being able to walk out my front door, shake hands with my customers, and see them off on another productive, no-hassle flight," he says. ♦

"CONVENIENCE AND SERVICE ARE EXTREMELY IMPORTANT, BUT WE PUT SAFETY ABOVE EVERYTHING ELSE."

▼ *With Executive AirShare, share-owners get to pick their flight times, choose the type of aircraft best suited to each trip, and have complete flexibility when it comes to return flights—all without the hassle or expense of owning a whole aircraft. The quiet, comfortable interiors turn flight time into productive work time.*

PHOTO BY ERIC FRANCIS

EXECUTIVE AIRSHARE | 231

CHAPTER THREE: ENJOYING GREATER KANSAS CITY

PHOTO BY MARIO MORGADO

PHOTO BY MARIO MORGADO

"Best sledding hill in the Kansas City area. Ask anyone in Kansas City who grew up there about Suicide Hill, and they will know where it is, and will probably have stories about it." That's the comment of Brent, a sledding aficionado and commenter on a Web site called Missouri Sled Riding Location, the existence of which indicates the gravity of sledding in Kansas City, Missouri. Suicide Hill has an approximate fifty-five-degree angle surface and offers a three- to four-hundred-foot icy ride. Suicide Hill is a designated sledding area in Brookside Park, topped by a tennis court with a baseball diamond at the bottom of the hill. The ride varies from fun to ferocious depending on where the sledders position their riding devices of choice. Like colorful penguins, children and adults repeatedly take their chances to see who can complete the most memorable slide on this typically very fast slope.

PHOTO BY DENNIS KEIM

PHOTO BY BRUCE MATHEWS

PHOTO BY DENNIS KEIM

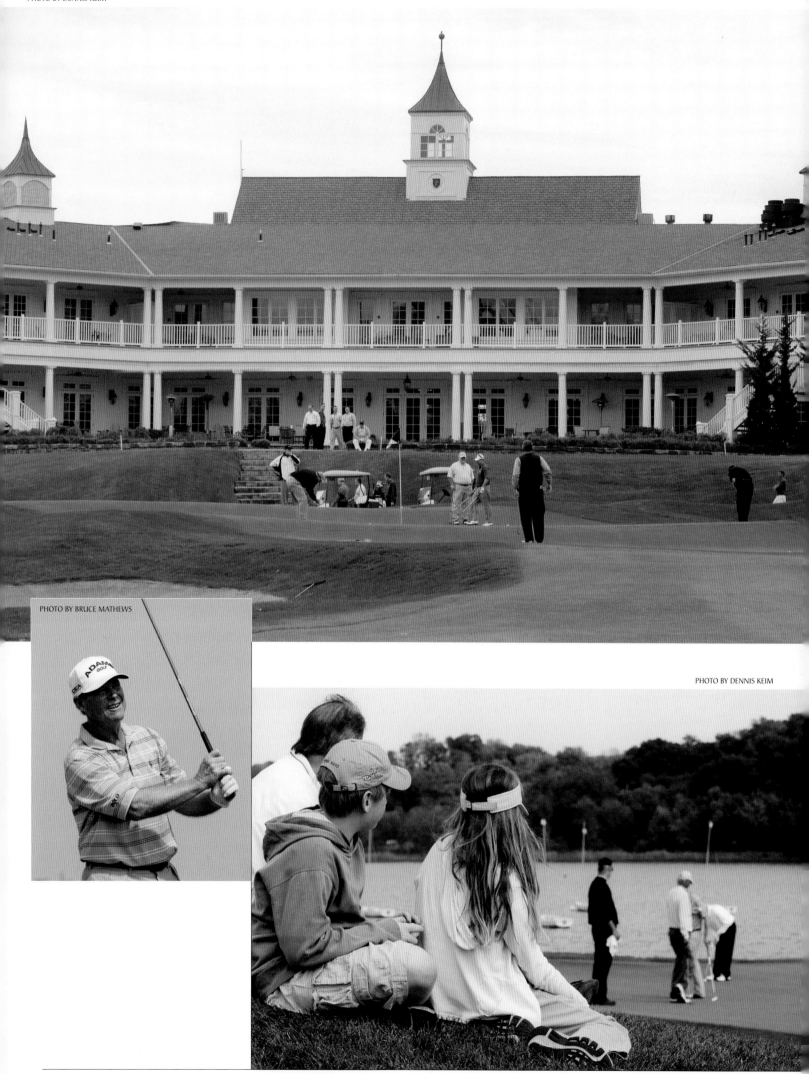

234 | GREATER KANSAS CITY: UNLIMITED POSSIBILITIES

CHAPTER THREE: ENJOYING GREATER KANSAS CITY

One of the advantages of living at the National Golf Club is access to not one but two 18-hole Tom Watson–designed golf courses. Members may even on occasion spot the legendary pro playing a round or two, as shown here. Born and raised in Kansas City, Missouri, Watson was ranked the number-one golfer in the world from 1978 to 1982. Today, when not playing in the occasional PGA Tour event, he lends his expertise to award-winning golf courses. In 2000, *Golf Digest* chose National Golf Club as one of the top-ten courses in Missouri. In addition to the golf course, the National Golf Club community features luxury homes nestled within rolling wooded terrain, a lake and docks, swimming pools, fitness centers, and a family activity center. Located just outside historic Parkville, the community is only fifteen minutes from downtown Kansas City.

Sprint: Setting the Standard in the Telecommunications Industry

The success of an industry leader like Sprint doesn't happen overnight. Instead, its success is the result of a long-term history of setting the highest standards for technological innovation and customer service excellence.

Sprint's history began in 1899 when founder Cleyson L. Brown organized the Brown Telephone Company in Abilene, Kansas. His competition, Bell, had a virtual monopoly on the market. But Brown persisted, and by the late 1920s his company had established operations in Kansas, Pennsylvania, Indiana, Ohio, and Illinois. By the 1950s, Brown's company had become the second-largest non-Bell telephone company in America.

Today, Sprint is guided by a single strategic vision: "To deliver the best products and services on the most powerful networks." Employees are encouraged not only to create innovative solutions to customer needs, but also to recognize opportunities for new products and services. The result has been a continual series of high-profile advances that have revolutionized the telecommunications industry. In 1986, for example, Sprint launched the nation's first coast-to-coast fiber-optic network as the centerpiece of its long-distance service. In 1996, it built the only nationwide PCS network in the United States. Its merger in 2005 with wireless communications company Nextel Communications will likewise ensure a steady stream of innovative, feature-rich, Internet-ready phones and smart devices.

▲ *This artist's rendering of Sprint Center (above) shows a venue that, when complete, will host a variety of sporting and entertainment events throughout the year. Currently under construction (right), Sprint Center is scheduled to open its doors in fall 2007.*

Today, Sprint continues its drive to set new standards of excellence in telecommunications. Its strategic focus is on the fastest-growing segments of the industry: wireless personal communications, the Internet, high-speed data, wireline and wireless broadband, and communications packages aimed at high-end consumers and companies. Standout products and services include two robust wireless networks, instant national and international walkie-talkie capabilities, and an award-winning and global Tier 1 Internet backbone.

No doubt Sprint is proud to lead the way in worldwide telecommunications advancements. The company is also proud to share the benefits of its leadership by taking a long-term interest in the communities in which its employees and customers work and live. Since its establishment in 1989, the Sprint Foundation has provided more than $93 million to community organizations across the country. In 2005, Sprint Foundation grants totaled $6.3 million, with nearly $1 million going to match employee contributions to youth development, arts and culture organizations, and education.

As part of its ongoing support of K–12 educational initiatives, Sprint established the Sprint Achievement Program in 2004, which has provided over $1 million to educators to implement projects in their classrooms that improve student engagement and increase family involvement. Additionally, Sprint is the Kansas City and Washington, D.C., sponsor of Connect with Kids, a multimedia character-education program focusing on the education and well-being of children.

Given its history in the Greater Kansas City area, Sprint is one of the region's strongest supporters, employing more than fourteen thousand people and committing

(continued on page 238)

SPRINT'S STRATEGIC VISION IS "TO DELIVER THE BEST PRODUCTS AND SERVICES ON THE MOST POWERFUL NETWORKS."

SPRINT

◀ *Sprint employees celebrate the refurbishing of a Kansas City resident's home during the annual citywide Christmas in October volunteer day.*

▼ *Sprint employees donate time, money, and food to the annual Harvesters' Food Drive. As the food bank serving Greater Kansas City, the Harvesters' network provides nearly sixty thousand people a week with emergency food assistance.*

(continued from page 237)

millions of dollars to Kansas City community outreach programs. Between 2004 and 2006 alone, over $20 million went to support various education, youth development, arts and culture, and community infrastructure projects. Sprint has also secured the naming rights and will provide all telecommunications services to downtown

CHAPTER THREE: ENJOYING GREATER KANSAS CITY

◀ Kansas City, Missouri, charter school Scuola Vita Nuova is one of many area schools to receive education-related grants from Sprint.

Kansas City's new Sprint Center. Set to open in 2007, Sprint Center will be the region's premier entertainment venue for sporting and entertainment events, and it will create the best possible fan experience by utilizing Sprint's innovative technology.

For its commitment to its employees, customers, and the community, Sprint has been the recipient of a number of prestigious awards. Key 2006 accomplishments include being named the U.S. Hispanic Chamber of Commerce's Corporation of the Year; being listed as one of Diversity-Inc's Top 50 Companies for Diversity (for the third consecutive year) and Top 10 Companies for African Americans; receiving the National PTA's Commitment to America's Children Award; and being inducted into the Mid-America Education Hall of Fame.

A company with an extraordinary heritage of achievement, Sprint continues to provide the products and services that enhance its customers' lives. By also focusing on the causes that are important to its employees and their communities, Sprint will no doubt remain the company of choice for all of its stakeholders. ♦

◀ The Sprint Campus in Overland Park, Kansas, is Sprint's operations headquarters and houses the largest percentage of Sprint Nextel employees.

SPRINT | 239

PHOTO BY SCOTT INDERMAUR

PHOTO BY SCOTT INDERMAUR

Used to be in the United States soccer was only played by schoolkids. Today, it's becoming a major professional sport. Major League Soccer was formed in the United States on December 17, 1993, as part of the US Soccer Federation's promise to the Fédération Internationale de Football Association (Europe's professional soccer association) to establish a Division One professional soccer league in exchange for hosting FIFA's 1994 World Cup. The Kansas City Wizards were one of the original ten teams organized as part of that promise. Since then they have boasted some of the league's best players (including #7 forward Eddie Johnson, shown at right), creating unprecedented enthusiasm for the game both nationally and locally. Here, local schoolkids perform in a pregame ceremony supporting the team in a home game against the New England Revolution. Home games are played in Kansas City's Arrowhead Stadium, which is also home to the NFL's Kansas City Chiefs.

CHAPTER THREE: ENJOYING GREATER KANSAS CITY

(a) The J. C. Nichols Memorial Fountain, located at the east entrance to Country Club Plaza, was created in the early 1900s by French sculptor Henri Greber and is adorned with four striking equestrian figures that are meant to represent the Mississippi, Volga, Seine, and Rhine rivers. (b) The Henry Wollman Bloch Memorial Fountain, erected in 2001, was gifted to the city in honor of the cofounder of H&R Block Inc. (c) Artist Wheeler Williams created the *Muse of the Missouri*, which is located on the Main Street median between Eighth and Ninth streets. (d) The Molamphy Memorial Fountain in Molamphy Memorial Park in Overland Park was erected in 1989 in honor of Richard Molamphy, the first president of the Overland Park Chamber of Commerce and an instrumental player in the development of College Boulevard. The sculpture by Overland Park artist Arlie Regier is entitled *Pierced Sky*. (e) Located in Penn Valley Park, the Firefighters Fountain is dedicated to the city's bravest who lost their lives in the line of duty. (f) The J. C. Nichols Memorial Fountain at night. (g) *The Spirit of Cooperation*, a fountain erected through contributions from both the public and private sectors, is one of the few fountains operating year-round. In winter, its cascades turn into intriguing ice sculptures. (h) The two sides of the fountain cascading from a brick wall on Eighth Street are distinctly different. Shown is Fountain II. (i) The Prairie Park Fountain, depicting a pioneer family, is located at the corner of Mission Road and Tomahawk in Prairie Village, Kansas.

CHAPTER THREE: ENJOYING GREATER KANSAS CITY

GREATER KANSAS CITY: UNLIMITED POSSIBILITIES | 243

PHOTO BY ERIC FRANCI

Hallmark Touches the Lives of Many

Throughout the past century, the Hallmark brand has come to symbolize the very best in creative expression, helping people share what is universal in the human heart. Simply put, Hallmark cards enrich lives through images and inspiration, sentiment and surprise, creativity and comfort.

Each year, Hallmark and its subsidiaries produce more than forty thousand products in over thirty languages distributed in one hundred countries. The Hallmark Channel reaches millions of subscribers in the United States with high-quality television programming. "Whenever, however, wherever people communicate, connect, and celebrate, Hallmark is there," says president and CEO Don Hall Jr., grandson of founder J. C. Hall.

The story of Hallmark could easily be a script for one of the heartwarming Hallmark Hall of Fame movies the public has come to enjoy. In 1910, with nearly empty pockets and nothing more than two shoeboxes of postcards under his arm, eighteen-year-old Joyce Clyde Hall began to fulfill his American dream by selling postcards to retailers throughout the Midwest. A brand—and ultimately, an industry—was born. His entrepreneurial spirit led him from postcards to greeting cards, spurred the invention of modern gift wrap, changed the way greeting cards were displayed in stores, created the renowned Hallmark Hall of Fame, and perhaps most important, made the company's name synonymous with quality and caring.

▲ *Eric Brace, master artist, both writes and illustrates for Hallmark's humorous alternative card lines Shoebox and Fresh Ink. To create greeting cards, he uses conventional media such as acrylic paints as well as digital tools.*

People in Kansas City know this. They also know that this American institution's mission of enriching lives is both broad and deep. To Hallmark, enriching lives also means enriching the communities in which it operates through the Hallmark Corporate Foundation's philanthropic support of children's health and education initiatives, citizen services, the arts, and urban communities.

Crown Center, one of the nation's first mixed-use real estate developments, is a shining example. The neighborhood around Hallmark's international headquarters was once a declining urban area defined by abandoned buildings, rutted parking lots, and litter. With the leadership of J. C. Hall and son Donald, redevelopment efforts started in the late 1960s to create Crown Center, with interconnected office buildings, two world-class hotels, expansive shopping and entertainment, and a residential community. Today, more than 5 million people work, live, and play at Crown Center each year, and Hallmark's long-standing commitment to the urban core stands as a model for other rejuvenation projects in Kansas City.

In keeping with the belief that enriching lives begins at home, Hallmark provides strong benefits, family-friendly work policies, and unparalleled resources for its vast creative team. The company also ensures that employees, known as "Hallmarkers," share in the success of the privately held company through profit sharing.

"It is an awesome responsibility, and an incredible privilege, to lead a company dedicated to enriching lives," Don Hall Jr. says. "At Hallmark, we are invited to give voice to people's feelings—of joy and grief, of compassion and healing. These are enduring human needs, which is why we have such confidence in the future of our company." ♦

HALLMARK CARDS ENRICH LIVES THROUGH IMAGES AND INSPIRATION, SENTIMENT AND SURPRISE, CREATIVITY AND COMFORT.

◀ *At the Hallmark Visitors Center, guests talk with a skilled craftsperson like press operator Rick Eisel as he demonstrates greeting card manufacturing processes such as hot foil stamping and die cutting.*

PHOTO BY ERIC FRANCIS

PHOTO BY ERIC FRANCIS

If you think John Wayne and a handful of cowboys won the West, think again. One visit to the American Royal Women's Ranch Rodeo National Championship will set you straight. Authentic cowgirls, right off the best working ranches in America's heartland, compete against each other roping and sorting cattle, doctoring livestock, and trailer loading. The action is fast paced, and the livestock are always unpredictable as these women demonstrate the skill and teamwork required to work on a real ranch. The American Royal is a not-for-profit organization that benefits youth and education, and which for more than one hundred years has celebrated the region's rich agriculture heritage through competition, education, and entertainment.

CHAPTER THREE: ENJOYING GREATER KANSAS CITY

Everyone knows that Kansas City is a city divided, part of it in Kansas, part of it in Missouri. That makes for some interesting rivalries, especially at the American Royal when the Young Riders Rodeo comes to Kemper Arena in the fall. The top high school and collegiate rodeo athletes show off their talents in breakaway roping, steer riding, barrel racing, and a real crowd favorite, mutton bustin' (usually for kids under fifty-five pounds, who ride a sheep out of a bucking chute and hold on as long as they can). The day before the Young Riders rodeo, contestants from Kansas and Missouri appear at the American Royal during the Youth Invitational Rodeo. After the competition is over, there's plenty of time to enjoy the excitement of the midway, where visitors can test themselves against the forces of gravity.

Kansas City Marriott Downtown: A Tradition of Hospitality at the Heart of the Renaissance

The Kansas City Marriott Downtown story is remarkably like that of Kansas City's. It has multiple layers that, when peeled back, reveal surprises that add intriguing texture to what appears at the surface.

A first glance reveals the city's largest convention hotel operating under the banner of an international mega-corporation recently hailed as one of the 400 Best Big Companies in America by *Forbes* magazine. The physical property encompasses two starkly contrasted landmark structures in the heart of the city's urban renaissance zone. One is a modern edifice built in the mid-1980s, the other an architectural treasure dating back to the early 1900s. An elevated walkway provides a physical connection between the two buildings.

But a deeper look reveals that they are connected in spirit by an extended family spanning two generations of members, whose lives are woven into the story of hospitality management in these landmarks for over half a century. In the balance of contrast is the soul of the Kansas City Marriott.

▲ *Located on the twentieth floor of the Marriott Tower, the executive lounge offers complimentary food and beverage presentations and breathtaking views of the dynamic downtown skyline.*

CHAPTER THREE: ENJOYING GREATER KANSAS CITY

PHOTO BY ALAN S. WEINER

"Our guests come with high expectations for Marriott consistency, and they receive that," says Carol Pecoraro, general manager. "The pleasant surprise comes with a unique brand of hospitality that flows from a culture steeped in independent innkeeping tradition. Our people make a positive difference in the lives of travelers in our care."

"Family" is at the essence of the Kansas City Marriott culture. "We believe in one another, and we stick together through good times and challenging times," says Pecoraro. Sticking together is a rare trait in an industry noted for high employee turnover, and yet, nearly one hundred staffers at the Marriott have worked together for more than ten years. Loyalty, dependability, work ethic, and a caring attitude are core values in this family. Pecoraro credits these values to the company's patriarch, Phillip Pistilli.

Pistilli, who began his career at the Muehlebach Hotel in 1954, was a legendary hotelier who left an indelible mark on the hospitality industry and on Kansas City. He developed and operated in the community some of America's most successful hotels.

"Mr. Pistilli was an innovator and a brilliant leader," says Pecoraro. "He expected hard work and loyalty, but he always returned what he got from us in greater measure." His egalitarian treatment of employees opened doors for women and minorities in management that were nonexistent at the time. "There was never a glass ceiling in his hotels," says Pecoraro. "Hard work and smarts equaled opportunity and reward. We continue to operate that way."

(continued on page 250)

"OUR GUESTS COME WITH HIGH EXPECTATIONS FOR MARRIOTT CONSISTENCY, AND THEY RECEIVE THAT."

▼ *Suite decor (as shown) is designed with a residential feel. Local landmarks are featured in prints of original oil paintings and photographs by a Kansas City artist.*

PHOTO BY DENNIS KEIM

(continued from page 249)

Another Pistilli legacy now bearing fruit was his belief in preservation and in downtown. Pecoraro, who guided the Marriott through a number of lean years, believes the company's patience and persistence are finally being rewarded. "There is tremendous energy downtown," says Pecoraro. "We are right at the heart of a historic renaissance. Development is at a level not seen since the heydays of the

▲ *The atrium with waterfall is a favored location for hotel guests to connect with one another.*

◀ *With nearly one hundred thousand square feet of meeting space, including thirty-four breakout rooms, the hotel can accommodate as few as ten or as many as two thousand meeting attendees.*

CHAPTER THREE: ENJOYING GREATER KANSAS CITY

◀ More than half a million local and hotel guests dine at the Kansas City Marriott each year. Many choose Lilly's, where custom omelets are a popular breakfast attraction.

original Muehlebach in the early 1920s and again in the post–World War II years."

The hotel is well-positioned for what promises to be a long-running economic growth cycle. "The Kansas City Marriott has the most architecturally diverse and interesting meeting space in all of Kansas City," she says. "With our newly renovated guest rooms, an expanded convention center, new office space, and attractions downtown, we can finally look forward to serving all three major market segments: conventioneers, business travelers, and pleasure travelers."

A major strength of the Kansas City Marriott is its banquet and meeting capabilities. With nearly one hundred thousand square feet of contemporary and historically significant function space, the hotel offers forty-two rooms of sufficient variety to create a uniquely suitable setting for an intimate executive conference, a prestigious social gathering, or a national convention. Architectural and decor distinctions facilitate event themes ranging from contemporary urban pop culture to classic jazz-age elegance.

The hotel is favored by local event planners for the city's most glamorous charity galas as well as corporate and civic functions. Of all the events produced at the Marriott, perhaps none receive more staff attention than weddings. "A wedding is the most important moment in life for a bridal couple and their families," says Pecoraro. "It is up to us to make a lifelong dream a wonderful, lasting memory." ♦

▼ The Imperial Ballroom in the Muehlebach Tower and the Count Basie Ballroom in the Marriott Tower are the perfect settings for elegant banquets of fifty to eighteen hundred attendees.

In Disney's *Geppetto & Son*, the classic story of the wooden boy Pinocchio is told from the point of view of the woodcarver who brings him to life. Scored by Academy Award–winning composer Stephen Schwartz (*Godspell, Wicked*), *Geppetto & Son* made its world premiere at the Coterie Theatre on June 27, 2006. The event was the third annual production in the Coterie's Lab for New Family Musicals, a program established at the theater to provide Broadway composers with a venue in which to showcase their works for young audiences. A nonprofit professional theater for youth and family audiences, the Coterie's main-stage season consists of six full productions from October to August, with an emphasis on recent works that target multigenerational families and young audiences aged five to eighteen. *Time* magazine named the Coterie one of the five best theatres for young audiences in the United States, and in 2004 the theatre won the Missouri Arts Award for Outstanding Contribution to the Arts.

CHAPTER THREE: ENJOYING GREATER KANSAS CITY

Eatin'—and Connectin'—Good in the Neighborhood

"We're not in the food business serving people, we're in the people business serving food."

That statement, expressed often by Applebee's International CEO and president Dave Goebel, sums up Applebee's philosophy. It's all about *people*—both the associates working in its restaurants and the guests it serves every day.

That people culture begins with the BIG Fun TRIP, an acronym for Applebee's principles and values (Balance, Innovation, Guest-driven, Fun, Teamwork, Results, Integrity, and Passion). The BIG Fun TRIP, quite simply, is Applebee's culture. Making that culture stick requires that everybody—including senior management—walk the walk, both with guests and with each other.

"Applebee's vision of becoming America's favorite neighbor is not just a marketing gimmick," said John Prutsman, executive director of human resources. "Applebee's invests in its people, who in turn invest in our guests and our communities. 'Neighborhood' is a sense of

▲ *Join Applebee's on the BIG Fun TRIP, an acronym that reminds associates about the company's principles and values. The company uses creative ways to bring its culture to life, including monthly spin-the-wheel contests to celebrate associates who epitomize a particular principle. Prizes have included one thousand dollars in cash.*

belonging, and at Applebee's it's at the heart of everything we do."

The philosophy obviously has generated success. Headquartered in Overland Park, Kansas, Applebee's Neighborhood Grill & Bar is the largest casual dining concept in America, with nineteen-hundred-plus restaurants in forty-nine states and seventeen countries.

Its community relationships are built one neighborhood at a time. Applebee's restaurants across the country generate programs at the local level with everything from flapjack fund-raisers to car washes, lemonade stands to Dining to Donate events. The events create a strong bond with guests and contribute to the communities as a whole. In addition, each restaurant reflects its neighborhood, with photographs of local high school sports teams, firefighters, and familiar landmarks adorning the walls.

That sort of community connection starts with Applebee's associates. "Hire great people, treat them as people first and employees second, and they'll make remarkable connections with your guests," Prutsman said.

Look no further than Applebee's Heidi Fund, established in 2002 in honor of Heidi Tomassi, a former Applebee's server. At the time, Heidi's four-month-old son was suffering from a rare heart defect. After two open-heart surgeries, Heidi and her husband were deeply in debt. However, when one of Heidi's customers accidentally left behind thirty-three hundred dollars in cash, she instinctively did the right thing and returned the money.

After the news broke, people nationwide donated fifteen thousand dollars, and Applebee's donated another ten thousand dollars. To honor Tomassi, Applebee's created the Heidi Fund to provide financial assistance to other Applebee's associates who find themselves in a financial crisis caused by a catastrophic event.

Associates can participate by making a payroll deduction of any size; no amount is too small. The company also donates annually, and since its inception, the Heidi Fund has provided a stunning $1.29 million in assistance to associates.

"Our associates' generosity never ceases to amaze me," said Jenny Truman, Applebee's charities coordinator. "We go the extra mile for one another, and we go the extra mile for our guests. It's what makes us great." ♦

PHOTO BY MARIO MORGADO

"WE'RE NOT IN THE FOOD BUSINESS SERVING PEOPLE, WE'RE IN THE PEOPLE BUSINESS SERVING FOOD."

◀ Since late 2002, Applebee's Heidi Fund has provided a total of $1.29 million to its associates, including cook Jeff Cavinaw when he underwent heart surgery. The program is funded largely by associates, who can donate as little as twenty-five cents a paycheck.

As they have done since it was first performed in 1904 at LaScala in Milan, audiences still thrill to the tragic love story of Navy Lieutenant B. F. Pinkerton and Cio-Cio-San, known as Butterfly. In a way, the steadfast faith exhibited in *Madame Butterfly* is mirrored in the long history of the Lyric Theatre and the Lyric Opera of Kansas City. Dedicated in 1926, the Lyric Theatre has gone through good times and bad. In 1942, the auditorium was stripped of its seats, sold to the American Red Cross, and used as a blood collection center. Subsequently it became a movie house. Eventually it was purchased by the Lyric Opera of Kansas City, repaired, renovated, and restored. Present-day Kansas City music lovers look forward each year to an ambitious season of operas such as *Madame Butterfly*. The Lyric also encourages new opera fans by offering innovative programs to further music and arts education in schools and in the community.

CHAPTER THREE: ENJOYING GREATER KANSAS CITY

PHOTO BY SCOTT INDERMAUR

GREATER KANSAS CITY: UNLIMITED POSSIBILITIES | 257

PHOTO BY SCOTT INDERMAU

Chase Suites Hotels:
An Extraordinary Stay for Business and Pleasure

A nationwide tradition in all-suite hospitality for over twenty years, Chase Suites Hotels exemplifies the "home-away-from-home" experience. As part of the Woodfin family of hotels, which were established in 1984 to accommodate the needs of corporate business travelers, Chase Suites Hotels specialize in providing upscale, all-suite accommodations and services to both business and leisure guests.

Visitors to Kansas City can choose from two convenient locations. The Chase Suites Hotel at Kansas City International Airport is located just four miles from the airport and is situated among North Executive Hills, a corporate office complex that is home to Farmland Foods, Worldspan, Toyota, ADT, Citicorp, American Airlines, and Harley-Davidson.

In town, the Overland Park Chase Suites Hotel at 110th and Lamar Avenue is located minutes from Sprint's World Headquarters Campus and Black & Veatch headquarters, across the street from the Overland Park Convention Center, and minutes from Town Center Plaza and Oak Park Mall.

Each hotel offers 112 contemporary and comfortable one- or two-room suites with twenty-eight penthouses, many at little more than the price of a regular hotel room. All accommodations include separate

▲ *Monday through Thursday evenings from 5:30 to 7 p.m., guests can enjoy appetizers or a light supper at the hotel's complimentary buffet. Manager-hosted, the spread includes a hot menu item, salad or veggie platter, cheese and crackers, and freshly baked cookies for dessert. Even the beer, wine, and soft drinks are on the house.*

living areas, fully equipped kitchens, comfortable beds, and spacious bathrooms. The two-room suites also include a fireplace.

To meet the high-tech needs of today's business travelers, each suite features high-speed Internet access and functional desk/workstations. Also available on-site are a twenty-four-hour business center with complimentary fax, copy, and PC services, and meeting facilities for up to fifty persons.

Designed to accommodate the needs of the traveling family as well, each suite offers free cable television and has VCR or DVD players available on request along with a selection of complimentary videos. And when Chase Suites says "Bring the family," they mean the *whole* family. The hotels' pet packages make furry family members feel right at home, with designer food and water bowls, designer pet bed, a personalized welcome from management, and a special treat upon check-in. Other hotel packages are also available, catering to frequent business travelers, honeymooners, sports fans, and holiday visitors.

Chase Suites also believes that because travel is expensive and time-consuming enough, guests are entitled to enjoy some free basic—and not-so-basic—amenities. Complimentary services include parking, pool and spa facilities, shuttle service, daily housekeeping, and even grocery shopping. If cooking isn't in the plan, each Chase Suites Hotel offers a free upgraded continental breakfast buffet, as well as evening social hour with drinks and a light dinner buffet, and is within walking distance of a number of restaurants.

Thanks to its long-standing tradition of blending comfort and convenience with style and affordability, Chase Suites Hotels experiences one of the highest customer satisfaction ratings in the hotel industry. In a recent survey, a full 94 percent said they would return to stay in a Chase Suites Hotel. Whether traveling for business or pleasure, alone or with the family, every guest who stays at Chase Suites Hotels in Kansas City is always made to feel right at home. ♦

EVERY GUEST WHO STAYS AT CHASE SUITES HOTELS IN KANSAS CITY IS ALWAYS MADE TO FEEL RIGHT AT HOME.

◀ *For the price of a regular hotel room, guests at Chase Suites Hotels stay in spacious luxury. All rooms are suite-sized—either Studio Queen, Studio Double, or the two-bedroom/two-bath Penthouse—making them the perfect home away from home for extended stays, families on vacation, and business travelers.*

PHOTO BY SCOTT INDERMAUR

Ready for a little time travel? About five hundred years and fifteen minutes from downtown Kansas City, you will be transported to a time of royals and peasants, knights and jesters, tavern wenches, blacksmiths, seamstresses, and strolling performers. After more than thirty years, the Kansas City Renaissance Festival has made a permanent home on sixteen acres in Bonner Springs. The company includes 500 costumed characters, more than 160 artisans, and performers on thirteen stages. For seven glorious weekends each fall, this festival provides an enchanting escape where knights duel on horseback to win the queen's favor, and visitors feast on food fit for a king. Of course, no trip would be complete without souvenirs, and craftspeople offer leather goods, jewelry, and glass trinkets. For truth seekers there are living history tours filled with intimate details about life in the shire.

CHAPTER THREE: ENJOYING GREATER KANSAS CITY

"Some people want to go to Vegas for fun," says Jaylene Lambert, owner of the Phoenix Piano Bar and Grill. "Others want to come to Kansas City and play jazz with Everett DeVan." Such are some people's passion, she says, that they fly in from all over the country just to sit in with the renowned musician and Kansas City native during his Tuesday night jam sessions. DeVan, shown here behind the mic, is just one of the many local greats who frequent the Phoenix. Well-known guitarist Matt Cooper is another. Located in a historic nineteenth century building downtown on Eighth and Central, the Phoenix first opened as a saloon in 1890. It has also been a famous 1940s supper club and, since 1990, a jazz club. Part of the fun, says Lambert, is you're never a stranger for long. "We usually fill every seat in the house. So if you don't know someone when you get here, you will by the time you leave."

CHAPTER THREE: ENJOYING GREATER KANSAS CITY

Bernie Kopell, renowned actor perhaps most remembered for his role as Dr. Adam Bricker on *The Love Boat* television series, took the stage for several months as part of the acting troupe at the New Theatre Restaurant. Kopell starred in the play *Leading Ladies* by Ken Ludwig, a comedy about two actors out to convince a dying millionaire they are her long-lost nieces. Year-round, the New Theatre Restaurant attracts celebrity performers to its dinner and matinee performances. Featuring five-star cuisine, the dinner theatre receives rave reviews for its plays, musicals, and comedy productions.

Sounds of Excitement: An Explosion of Revitalization Resonates in the Kansas City Symphony

Amid the upsurge of development in downtown Kansas City is an organization experiencing an exciting transformation all its own. This is the Kansas City Symphony, a musical entity turned cultural icon for one of the nation's most exciting, distinctive metropolitan areas.

"The growth and development of the Kansas City Symphony mirrors that of the city itself," says Frank Byrne, executive director. "Kansas City has defined for itself a vision of being a world-class city, and no great city in the world exists without a great symphony orchestra."

Founded in 1982 by R. Crosby Kemper Jr., the Kansas City Symphony has become an organization of eighty musicians whose artistic talents serve a population of nearly 2 million residents.

Over its forty-two-week season, the symphony performs in a variety of formats and venues. In addition to classical, pops,

▲ *Music director Michael Stern leads the Kansas City Symphony in concert before a capacity audience. Stern's dynamic musicianship, compelling performances, and artistic vision have made the symphony an even greater asset for the city and the region.*

CHAPTER THREE: ENJOYING GREATER KANSAS CITY

PHOTO BY SCOTT INDERMAUR

In fact, a supportive community is one of the factors behind the symphony's momentum. "People see that we really care about sharing great music with them, and when we reach out, the community responds," says Byrne. That response meant sellout concerts and record attendance for the symphony during the 2006 season, including five of the most well-attended performances in its quarter-century history.

As a result, the symphony is experiencing an exceptional fiscal soundness, marked by an operating budget exceeding $10 million annually and an endowment that has surpassed the $35 million mark.

Much of the symphony's leap into the cultural spotlight began in 2004 following the appointment of music director Michael Stern, widely recognized as one of the most dynamic conductors of his generation. Stern's impressive musical talent, combined with a visionary governing board comprising civic and corporate leaders, continues to guide the symphony as it builds on its mission to enrich the community through the beauty and power of symphonic music.

(continued on page 266)

"THE ART OF LIVE PERFORMANCE IS ABOUT CONNECTING WITH PEOPLE IN REAL TIME."

▼ *Reaching out to the community is a key mission for the Kansas City Symphony. Free concerts for the public are among the many family-friendly events that the symphony provides around the metropolitan area that attract thousands of attendees each year.*

education, and family concert series, the symphony performs for the Lyric Opera of Kansas City and the Kansas City Ballet.

The symphony also takes its music to the community through educational programs, ensemble presentations, and free public concerts. "The art of live performance is about connecting with people in real time, experiencing together the incredible discipline and virtuosity that allows eighty musicians to perform as one," explains Byrne. "That connection is one that can occur in settings both large and small, and it provides each person with a unique experience."

PHOTO BY THOMAS S. ENGLAND

KANSAS CITY SYMPHONY

PHOTO BY SCOTT INDERMAUR

◀ *Grammy Award–winning singer Amy Grant appeared with the symphony for its highly successful Pops Series. The Kansas City Symphony's Pops concerts feature legendary performers, music of Broadway, and contemporary artists, all accompanied by a full symphony orchestra.*

(continued from page 265)

The symphony's future will be defined and enhanced by its new home in the Kauffman Center for the Performing Arts. This world-class center will provide the Kansas City Symphony with new opportunities to transform itself and, in turn, the concert-going experience, allowing the symphony to strengthen its communal bonds and to continue its efforts to inspire and entertain the people of this exciting midwestern region. ◆

PHOTO BY THOMAS S. ENGLAND

◀ *The Kansas City Symphony brings world-class guest artists and conductors to the stage in its classical series, which is at the heart of the symphony's mission to present great music for the community.*

CHAPTER THREE: ENJOYING GREATER KANSAS CITY

Like the setting sun, the rivers flowing through Kansas City are vital to life in this metropolis. In addition to being a mode of transport for cargo in and out of the region, the rivers serve as a primary water source for area residents.

PHOTO BY ERIC FRANCIS

From silly to a little scary, StoneLion Puppet Theatre creates a magical world of fun and fantasy for audiences around the region. Using marionettes, masks, and a host of enthralling puppetry styles, the troupe performs for private groups or in public venues to audiences of all ages and interests.

PHOTO BY ERIC FRANCIS

CHAPTER THREE: ENJOYING GREATER KANSAS CITY

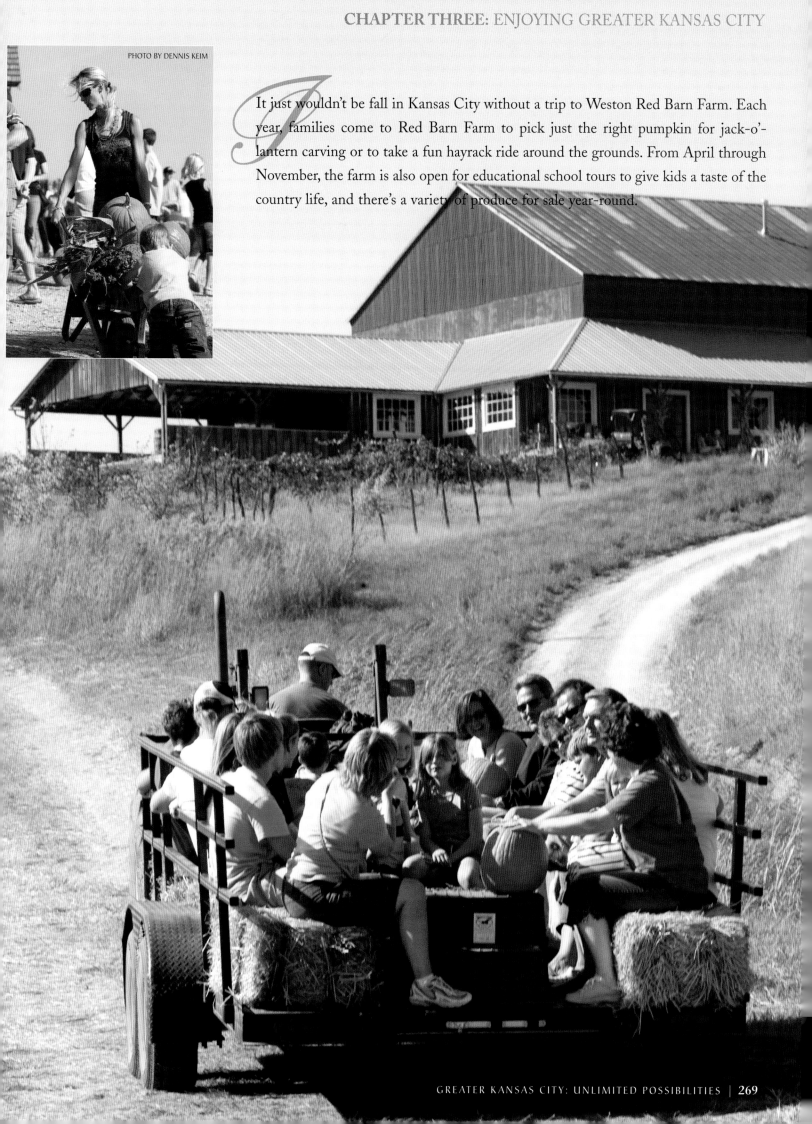

It just wouldn't be fall in Kansas City without a trip to Weston Red Barn Farm. Each year, families come to Red Barn Farm to pick just the right pumpkin for jack-o'-lantern carving or to take a fun hayrack ride around the grounds. From April through November, the farm is also open for educational school tours to give kids a taste of the country life, and there's a variety of produce for sale year-round.

Discover Gracious, Old World Charm at Hotel Phillips

Kansas City in the early 1930s was very much the crossroads of the United States, bursting with economic energy and musical creativity. When it first opened in 1931, the Hotel Phillips reflected this jazzy spirit. At twenty stories high, Hotel Phillips was the city's tallest hotel. It was also its most luxurious, an art deco masterpiece of elegance and first-class service that catered to clientele that included U.S. servicemen, celebrities, and presidents Eisenhower and Truman.

Placed on the National Register of Historic Places in 1979, in 2001 Marcus Hotels & Resorts reopened the property after a spectacular $20 million renovation. Today, Hotel Phillips blends vintage opulence and hip, contemporary luxury, while still delivering the AAA Four Diamond Award service that makes it a member of the prestigious Preferred Hotels collection.

Each of the hotel's 217 guest rooms and suites offers the amenities expected by both business and pleasure travelers. Accommodations include fine linens, deluxe baths, evening turndown, and concierge and valet services with same-day laundry and dry cleaning. Free access to

▲ *One of downtown Kansas City's most beautiful art deco buildings, Hotel Phillips transports its guests back to an era when all guests were considered royalty—and it still treats them that way. Rich in period detail, as shown by this view from the mezzanine into the lobby, Hotel Phillips mixes old-time glamour with modern luxury and convenience. Presidents, movie stars, and other dignitaries once graced these interiors. Today, the hotel is a favorite of pleasure and business travelers as well as those seeking an elegant, full-service hotel for their meetings and special events.*

CHAPTER THREE: ENJOYING GREATER KANSAS CITY

PHOTO BY ALAN S. WEINER

as well as the Truman Sports Complex and the Nelson-Atkins Art Museum. Once the city's planned $300 million performing arts center is built, Hotel Phillips will be squarely located within the city's arts community.

Guests are also delighted to discover that one of the most notable restaurants in town is located right on-site. The Phillips ChopHouse is consistently rated one of the best restaurants in Kansas City for its sophisticated atmosphere and inspired cuisine. There's even a private dining room that can accommodate meetings or special events of up to twenty-two people. The hotel's more casual eatery, 12 Baltimore, is the perfect spot for a relaxed business lunch or after-hours martini. The locals also know it's a great place to catch live jazz, swing, and zydeco music on Friday and Saturday nights.

As a premier meeting destination, Hotel Phillips specializes in smaller executive meetings and business events. With

(continued on page 272)

cable television and high-speed Internet, and two-line phones are also part of the room rates. Additional amenities include a fitness center, massage treatments, an in-house florist, and Wi-Fi in the lobby and meeting rooms.

Conveniently located in the heart of downtown, Hotel Phillips makes a great base from which to explore the city. Just a short ride from the Kansas City International Airport, the hotel is also located two blocks from the Kansas City Convention Center and numerous attractions, such as fine shopping and dining, museums, the city's historic jazz district, the Crown Center and Country Club Plaza,

HOTEL PHILLIPS REMAINS THE PLACE TO STAY IN DOWNTOWN KANSAS CITY.

▼ *Open for breakfast, lunch, and dinner seven days a week, the hotel's bar and grill, 12 Baltimore, features an extensive libations menu and live music on Friday and Saturday nights.*

PHOTO BY MARIO MORGADO

PHOTO BY ERIC FRANCIS

(continued from page 271)

fifty-six hundred square feet of meeting space and nine rooms, the hotel can accommodate up to 275 people at one time. Additional services include videoconferencing, a full-service business center, and catering. Even better, it's all offered in the ambience of a European-style boutique hotel for those guests who appreciate a dash of elegant form along with their function.

It's no surprise that Hotel Phillips is a top choice for social events like weddings, anniversaries, and private parties. For a truly special evening, nothing compares to the Crystal Ballroom, with its deco embellishments, sparkling chandeliers, and hardwood floors. The hotel's majestic mezzanine is also a great spot for cocktails, hors d'oeuvres, or dinner.

The hotel is gaining in popularity as a venue for destination weddings. Its wedding coordinator and support staff specialize in

▲ *The hotel's Phillips ChopHouse is one of the most consistently excellent restaurants in Kansas City. Featuring an award-winning seasonal menu, it was also presented with a* Wine Spectator *magazine Best Of award in 2006.*

◄ *For over seventy-five years, Dawn has greeted guests as they ascend the lobby stairs up to the hotel's second floor. Standing eleven feet high, the golden personification of the winged goddess of the stars was sculpted by Kansas City artist Jorgen C. Dreyer.*

CHAPTER THREE: ENJOYING GREATER KANSAS CITY

◀ Original to the hotel, the elegant Crystal Ballroom is so named because of its sparkling chandeliers. They're just one of the elements that make the ballroom a favorite spot for special events—everything from large wedding receptions to intimate dinners for two.

PHOTO BY MARIO MORGADO

creating not only a memorable wedding day, but also a memorable stay for the duration of the event. Everything is arranged on-site, from the bridal lunch to rehearsal dinner to tours of downtown's sights for the guests. The special day itself will be coordinated with the hotel's banquet staff, in-house sommelier, and floral designer to fit each couple's individual budget and taste.

Built as a showcase for a bustling city, the Hotel Phillips retains its status as one of downtown Kansas City's most charming and beautiful boutique hotels. Seamlessly blending old-world style and gracious service with modern-day amenities and business technologies, Hotel Phillips remains the place to stay in downtown Kansas City. ♦

◀ Riding the crest of a wave, her arms stretched overhead as she presents a lit torch skyward, Dawn is the symbol of the hotel's hospitality. Although a mythical figure representing sculptor Jorgen Dreyer's ideal woman, he used as his models several very real women, including Virginia Raupp, who at the time was a fifteen-year-old high school student from Lee's Summit.

a. The Nelson Sculpture Garden at dusk.

b. Upright Motive No. 9 by Henry Moore.

c. Shuttlecocks by Claes Oldenburg and Coosje van Bruggen.

d. Large Interior Form by Henry Moore.

e. The Thinker by Auguste Rodin.

f. Seated Woman by Henry Moore.

The Kansas City Sculpture Park's twenty-two acres, designed by architects Dan Kiley and Jaquelin Robertson, hold over thirty sculptures, including a cast of The Thinker by Auguste Rodin. The Nelson-Atkins Museum of Art, recognized as one of the nation's finest overall collections, boasts more than thirty-four thousand works. Among the museum's pieces are the world's largest collection of Henry Moore works, including Upright Motive No. 9, which was inspired by the totem poles carved by the Native Americans of the American Northwest. Seated Woman is another sculpture by Henry Moore, as is Large Interior Form, an abstract of the human form created by the artist. In 2007, the museum celebrated the opening of the Bloch Building, an addition of architectural artistry, designed by Steven Holl. The building, which is primarily underground, features five glass "lenses" emerging from the structure that reflect surrounding elements and create a changing visual display. This 840-foot addition is one of the highlights of an extensive $200 million restoration of the museum and sculpture park.

CHAPTER THREE: ENJOYING GREATER KANSAS CITY

CHAPTER THREE: ENJOYING GREATER KANSAS CITY

PHOTO BY MARIO MORGADO

Once blighted, the Crown Center area in downtown Kansas City, Missouri, was revitalized by Joyce C. Hall, Hallmark Cards Inc. founder, and his son Donald J. Hall, current Hallmark chairman. Now vibrant with businesses, luxury hotels, and a six-acre residential community, the area steadily attracts visitors with entertainment facilities such as the Crown Center Ice Terrace. The thirty-four-year-old open-air skating arena is available from November to mid-March, and provides skating lessons for students as young as three years old. "On the day after Thanksgiving, we hold the Mayor's Christmas Tree Lighting here, drawing anywhere from ten thousand to twenty thousand people," said Brett Purcell, manager. When the terrace closes for the season, the square remains alive with festivals and events, but fans of the terrace (some fifty thousand skaters per year) never fail to arrive when the rink is open again for a cool time on the ice.

▲ *The MAX, a rapid-transit bus service connecting popular downtown destinations, has become a model for cities nationwide. The line has already proven to be an attractive option for riders who might have other forms of transportation and has become a vital part of the city's urban core.*

HNTB: Building a Better Tomorrow

As a multidisciplinary architectural, engineering, and planning firm whose portfolio includes some of Kansas City's most distinctive landmarks and progressive projects, HNTB knows how to make the world a better place. "Everything we do as a business is geared toward promoting quality of life, economic development, and sustainability of our infrastructure," explains HNTB Corporation president Scott Smith, P.E.

While its work ranges from plazas and parks to airports and expressways to buildings and bridges, every project, regardless of size, is one of distinctive quality, delivered on time, within budget, and to each client's satisfaction. "As a client-focused firm, we gear everything toward helping our clients solve problems and be more successful," says Smith, adding that everyone, on some level, is a client. "Our work has an aspect of the 'designer as user' approach. We always do our very best design because we'll be using these structures ourselves for the next fifty years."

With its national headquarters in Kansas City, HNTB oversees offices across the country. In Kansas City alone, there are numerous examples of HNTB's ability to provide viable solutions.

One is the renovation and expansion of the Kansas City Convention Center. When faced with space limitations, HNTB looked skyward and found the answer in the air

CHAPTER THREE: ENJOYING GREATER KANSAS CITY

PHOTO BY SCOTT INDERMAUR

Another impressive example of HNTB's work is the Kansas City Area Transportation Authority's (KCATA) Metro Area Express Bus Rapid Transit design. The MAX line, now a model for other cities, connects the city's popular downtown destinations and is designed to be easy-to-use, reliable, and attractive to riders who might choose it over private transportation. "Part of our effort during development was to provide public education that transit is a vital component of a lively urban core, that it really enhances the urban environment," says Wayne Feuerborn, HNTB's Kansas City director of urban design and planning. "Today, municipalities and communities across the country are recognizing the potential benefit of a similar system."

In addition to importing work from every coast, HNTB's sterling reputation attracts the nation's best design professionals to Kansas City. Through its work nationwide, HNTB is a major contributor to Kansas City's economy and to its future as a global competitor. ♦

PHOTO BY ALISE O'BRIEN PHOTOGRAPHY

space over the adjacent interstate, creating a 46,500-square-foot ballroom that spans seventeen lanes of traffic.

"Our work on the $150 million renovation/expansion of the Kansas City Convention Center is intended to reposition the center in its marketplace for the next ten to fifteen years," says Todd Achelpohl, principal architect. "The updates are a response to market demands for technology features and sophisticated meeting environments that are now the standard for this competitive industry."

With plans for the new ballroom to be certified as a Leadership in Energy and Environmental Design "green" building, the project places the city squarely among the nation's top convention spots.

"WE ALWAYS DO OUR VERY BEST DESIGN BECAUSE WE'LL BE USING THESE STRUCTURES OURSELVES FOR THE NEXT FIFTY YEARS."

◀ As part of the $150 million renovation of Kansas City Convention Center, HNTB nearly doubled the facility's space by creating an expansion that spans seventeen lanes of traffic. The reconfigured interior now features more than twenty meeting rooms and sophisticated, energy, and cost-efficient lighting that enhances any event.

Since moving to Kansas City in 1963, the Kansas City Chiefs have been a major player in the area's sports offerings. After local business leaders fulfilled their promise to sell enough tickets to fill a stadium, the team then known as the Texans was brought to Kansas City by Lamar Hunt, one of the founders of the American Football League and owner of the team until his death in 2006. With the move, the team was named after Kansas City mayor H. Roe Bartle, who earned the nickname "Chief" while working with the Boy Scouts of America. As a young team, the Chiefs became a contender in the first Super Bowl ever played, and took the title three seasons later in a win over the Minnesota Vikings with a score of 23-7. In 1971, the team played the league's longest game, going eighty-two minutes in double overtime against the Miami Dolphins. Under the coaching of Martin Edward "Marty" Schottenheimer, the Chiefs would later head to the playoffs for six straight seasons, taking three of the team's eight division titles it has earned over time. The Chiefs tout one of the National Football League's most spirited fan bases, who cheer their team on in Arrowhead Stadium.

CHAPTER THREE: ENJOYING GREATER KANSAS CITY

Car racing has come a long way from its early beginnings, when drivers pushed the speed limits of the first internal combustion engines on the sand tracks at Daytona Beach, Florida. Today, NASCAR has grown into the largest sanctioning body of motor sports in the United States, featuring some of the world's best drivers and automobiles, and an incredibly dedicated fan base. In fact, it's often difficult to tell which is louder—the roar from the stands or from the racing engines. Since 1999, the Kansas City Speedway has been an important venue on the racing circuit. In fact, on a NASCAR Nextel Cup Series race weekend, Kansas Speedway becomes the fourth-largest city in Kansas. Featuring a 1.5-mile tri-oval suitable for all types of racing and an eighty-two-thousand-seat grandstand, the speedway is located at the intersection of I-435 and I-70 in Kansas City, Kansas, about fifteen miles west of downtown.

◀ Becky S. Wilson, president of WDS Marketing & Public Relations, stands by the gigantic book fronts featuring favorite local authors at the Downtown Branch of the Kansas City Public Library. Wilson founded her agency in 1991 in Mission, Kansas.

PHOTO BY THOMAS S. ENGLAND

WDS Marketing & Public Relations Excels in Building Brand and Awareness

At WDS Marketing & Public Relations, the goal is for continuous improvement—to provide more value and increasingly superior results for clients. President Becky Wilson and her team hone areas of expertise to develop customized, integrated marketing and public relations programs.

Services at WDS include business development, media relations, brand building, business milestone marketing, and business-themed events. The agency, active in local and regional business since 1991, has enjoyed great success in obtaining media coverage in local and national business magazines, industry publications, and newspapers. Clients include dynamic mid-sized and entrepreneurial businesses in all areas of commerce.

WDS excels in assisting clients with participation in awards programs. "This type of brand building is an ideal venue for creating greater awareness and building credibility," according to Wilson. With help from WDS, clients have received awards such as Top 10 Small Businesses of the Year; the Business Ethics Award, the Women Who Mean Business, and 25 under 25 small business honors; and many others.

"Wilson's agency assisted us in seeking a highly coveted small business award. They delivered outstanding insight and service, including guiding us through the winner's celebration activities and events," says Michael D. Fuhrman, vice president of Schutte Lumber Company. "We enjoyed the entire experience."

The Protégé for a Day entrepreneur auction program is WDS's signature event, created for the KC Council of Women Business Owners. WDS was awarded with the Business Marketing Association's highest honor, the Fountain Award, in 2005 and 2006. ♦

WDS EXCELS IN PROVIDING CLIENTS WITH CUSTOMIZED MARKETING AND PUBLIC RELATIONS PROGRAMS.

CHAPTER THREE: ENJOYING GREATER KANSAS CITY

Located at one of the country's jazz crossroads—18th & Vine in Kansas City, Missouri—the American Jazz Museum celebrates all things jazz, past and present. New Orleans is often called the birthplace of jazz, but Kansas City is where jazz grew up. The city was christened "Paris on the Plains" during Prohibition (1919–33), when Tom Pendergast, mob boss, allowed the free flow of alcohol accompanied by all-night music. Musicians, many known today as the legends of jazz, flocked to the area, and their improvisational talents inspired swing and bebop music. During the day, the Blue Room, part of the museum, offers exhibits and one of the largest collections of jazz videos, but at night it transforms into a nationally acclaimed working jazz club. Museum patrons can listen to recordings and career information about jazz greats at kiosk listening stations, view collections and memorabilia, and mix music in the on-site studio.

PHOTO BY THOMAS S. ENGLAND

PHOTO BY THOMAS S. ENGLAND

PHOTO BY THOMAS S. ENGLAND

Kansas City is legendary for blues and jazz—not just for musicians like Count Basie and Charlie Parker, who both began their careers here, but for the current roster of headliners and local talent looking to make their mark. Each year the Kansas City Blues & Jazz Festival draws upward of thirty thousand spectators for the two-day event. Most recently held on Labor Day weekend at the Woodlands Race Track, the venue provides ample covered grandstands to shelter fans from the sun and elements. With the possible exception of steaks and barbecue, Kansas City is most noted for a distinctive jazz musical style, which consists of a two-four beat, the predominance of saxophones, and background riffs. Fans can see Charlie Parker's sax at the American Jazz Museum and dance in the Blue Room, which *Downbeat* magazine recognized as one of the one hundred greatest jazz clubs in the world.

PHOTO BY DENNIS KEIM

PHOTO BY DENNIS KEIM

CHAPTER THREE: ENJOYING GREATER KANSAS CITY

What can be more exciting or nostalgic than a homecoming football game on a crisp fall night? North Kansas City High School (NKCHS) in North Kansas City, Missouri, celebrated the event for an entire week prior to the game with local parades, parties, and school assemblies. Students and alumni (including a large group of 1951 graduates) cheered the school team, the Hornets (in purple and gold), as they went head-to-head against rival Truman High School. Waiting anxiously for the halftime crowning of the queen are homecoming candidates (left to right) Rachel McCommon, Lindsay Holland, Christina Ballance, Kelsey Strange, Ellie Brewer, Sam Kapp, and Brittney St. John. Strange won the crown, the Hornets won the game, and the evening culminated with a romantic "Egyptian Nights" theme dance. NKCHS currently enrolls about eighteen hundred students (grades 9–12) and, built in 1925, is the oldest district high school.

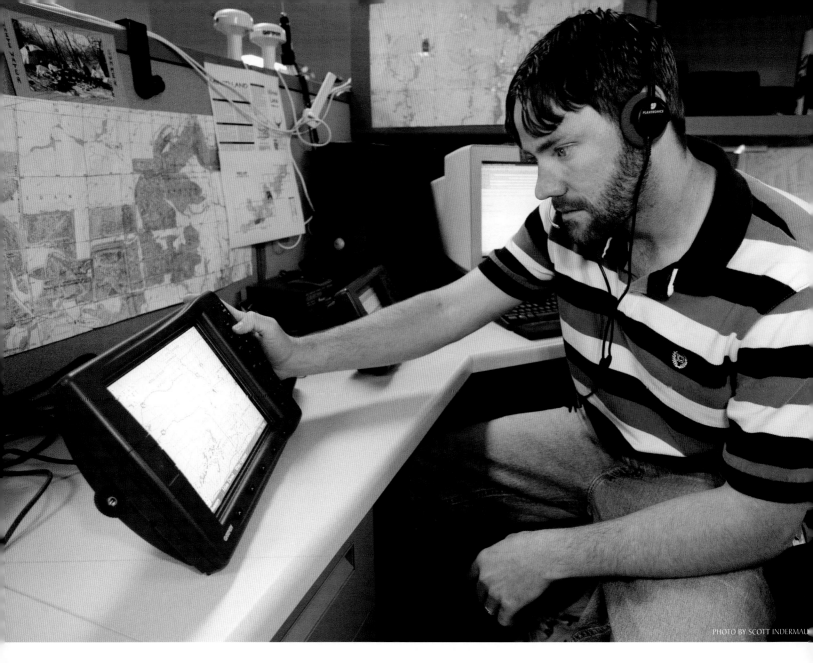

You Know Where You Are with Garmin

The GPS coordinates are N 38 deg 51.333 min, W 094 deg 47.941 min. The location is East 151st Street, Olathe, Kansas. The company is Garmin International, Inc.

In 1989, Gary Burrell and Min Kao had a great idea for a product that took advantage of the Global Positioning System (GPS). Today, Garmin (the Garmin name is a combination of the founders' first names) has a diverse product line and distributors in virtually every part of the world with products serving general aviation, marine, automotive, outdoor, mobile electronics, and personal fitness. The Garmin brand is known for innovative, award-winning products that set the standard in the industry.

"We remain committed to the creation of innovative and feature-rich products," said Dr. Min Kao, Garmin's president and CEO. Garmin currently owns 291 patents, and the product line is constantly increasing. In 2006 alone, it announced approximately seventy new products.

With offices in the United States, Europe, and Asia, Garmin has a product that fits every lifestyle and budget. Many people first encountered GPS technology

▲ *Garmin offers superior products supported by superior customer service. Answers to many questions are located on their Web site, but if customers need further assistance, they can call a hotline to put them in touch with experts who know the product inside and out.*

in automobiles. In fact, if the last car you rented had a portable GPS unit, chances are it was manufactured by Garmin.

For instance, the nüvi™ series integrates voice-prompted, turn-by-turn directions with a personal travel assistant that gives language translations, travel guide books, Bluetooth, and much more. Of their Garmin nüvi, one customer said, "Since I've owned it, I haven't been lost one time and I drive twenty-five thousand miles a year."

Then there's the Rino®, a rugged, two-way radio with GPS that is a real favorite with hikers and climbers. The Rino's patented location-reporting feature allows you to send and receive GPS positions with other Rino users.

For those interested in fitness, Garmin practically invented the GPS fitness category to provide athletes with information like time, speed, distance, calories burned, and heart rate. One customer said it well, "I chose the Forerunner 305 because it is the best training product, providing a great deal of data in flexible formats designed to support almost any training program."

In addition to products, Garmin has also launched another innovation. In late 2006, the company opened its first retail store, located on Chicago's Magnificent Mile on Michigan Avenue. It is the first store in the world dedicated to GPS devices. "As the GPS market continues to expand, we continue to build our leadership position through brand-building efforts that directly touch our customers. We are thrilled with the opportunity to directly interact with customers at our flagship retail store," said Kao. Unlike a traditional electronics store where GPS devices are locked under glass, the Garmin store is hands-on so customers can test the products.

Garmin's immediate success comes from developing innovative products, but their long-term success is from employees. The Garmin team will tell you they like to laugh, have fun, and work hard. Garmin has a unique business structure based on performing design, manufacturing, and marketing processes all in-house, which allows them to maintain a high level of quality and be more responsive to the everchanging marketplace. Garmin's hands-on approach equals products that are reliable and easy to use—just what the customers ordered.

Garmin engineers can be found at their workbenches creating, designing, and evaluating a product from start to finish, thus allowing the engineers to play a huge role in the product. A few months after their creations are complete, they can see the product on store shelves.

Another Garmin hallmark is customer support. All support is done in-house.

(continued on page 290)

"WE REMAIN COMMITTED TO THE CREATION OF INNOVATIVE AND FEATURE-RICH PRODUCTS."

▼ *People living in the Kansas City area recognize this as the Garmin International building in Olathe. Garmin started with a handful of dedicated engineers who had a great idea, and the company has since become the GPS leader both in sales and critical acclaim.*

PHOTO BY SCOTT INDERMAUR

GARMIN INTERNATIONAL, INC.

PHOTO BY SCOTT INDERMAUR

(continued from page 289)

When someone calls looking for help, they talk to *Garmin employees* at Garmin's facility who know the products inside and out. In many cases, employees have used the products themselves. Garmin even has a system in place where employees can check

▲ *Garmin has established many "firsts," from landing the first nonprecision approach using a general aviation GPS and creating the first palm-sized GPS/VHF, to introducing for outdoor enthusiasts the world's smallest GPS. Engineers working on prototypes have discovered that if they take care of cutting-edge technology, the "firsts" seem to take care of themselves.*

◀ *Whether you are on the road, on the trail, on the water, in the air, or even into fitness, Garmin GPS units are there too. Technicians build the units that are known for their simple operation, logical menus, smart features, and long-lasting reliability.*

out products so they can become more familiar with the technology.

If all this sounds exciting, it's just part of Garmin's creative, supportive culture. The family atmosphere carries over to the annual holiday party for friends and family and the summer event at Worlds of Fun. The areas where Garmin has locations also profit from that family spirit. The company and the employees take an active role in giving back to their communities by supporting agencies such as the United Way, the Girl Scouts and Boy Scouts, American Heart Association, Salvation Army, and YMCA, just to name a few.

So what is it like to be part of the Garmin family? One employee summed it up well. "Garmin's fun, it has great benefits, and we get to build cool electronics." ♦

PHOTO BY SCOTT INDERMAUR

◀ Technology breakthroughs at Garmin have enabled the company to develop GPS receivers that locate satellite signals quickly, maintain a lock on those signals, track users' locations wherever they go, and extend the life of the receivers' batteries.

▼ Regardless of where in North America you purchased your new Garmin GPS, you can be sure it went through their warehouse and distribution center in Olathe, Kansas. The 125,000-square-foot warehouse is the home of Garmin's North American distribution. Over two miles of conveyor belts increase efficiency and help ensure that Garmin products reach their destinations on time.

PHOTO BY MARIO MORGADO

Uptown

BATTLE OF THE BANDS
NOV 25
DOORS 5 PM

PHOTO BY MARIO MORGADO

Highly supportive of its local music scene, Greater Kansas City is home not only to great venues, but also to top-notch promotional organizations. One of them, the Web-based promotional site Bands Across Kansas City, sponsors a number of events like the Local Music Playoffs, designed to recognize and promote Kansas City's musical talent. Every genre of music is represented in this yearly event—from hip hop to alternative to metal—and there's even a special segment devoted entirely to high school bands. In 2006, the final night of the four-day playoffs was held at the historic Uptown Theater and featured popular alternative/pop rock band the Brisbanes, whose guitarist is shown here. A top entertainment venue since opening in 1928, the Uptown is regarded as one of the best live music venues in the city.

CHAPTER THREE: ENJOYING GREATER KANSAS CITY

"That was horrifying!" is a great review for "The Beast," the legendary haunted house that pioneered wandering through a maze of frightening surprises—known as the "open format." Seasonally operating since 1991, a scary journey through "The Beast" takes forty minutes (if you don't disappear). And disappearing may be possible since the werewolf forest alone covers one-fourth of an acre. Even the escape is blood-curdling—a straight-down, speedy, four-story slide to the exit. Visitors to Kansas City, Missouri, named the "haunted house capital of the world" by *PM Magazine*, can also say they've been to the "Edge of Hell," the oldest of the city's haunted abodes. The converted five-story warehouse contains a walk-through stocked with forty-five live performances, twenty-three terrorizing scenes, a live anaconda, and a five-story slide into the arms of the devil. Full Moon Productions, the owners and operators, beckon you to enter the houses every Halloween season—if you dare.

Kessinger/Hunter & Company: Putting Clients First since 1879

Success in the commercial real estate industry isn't just about leasing the right space or the right building. It's also about vision. Some companies may see things only as they are, but Kessinger/Hunter & Company sees things as they could be.

Take the former Sunflower Army Ammunition plant, for example. Once the economic engine that drove the City of De Soto, by the late 1980s it had fallen into ruin. Worse yet, it was an environmental hazard. But Kessinger/Hunter didn't see dilapidation. It saw potential. In 2005, the company partnered with Denver-based International Risk Group to clean up and redevelop the nine-thousand-acre site into homes, schools, churches, businesses, research facilities, parks, and open space.

"The process of dealing successfully with multiple government agencies, private contractors, and legal hurdles demonstrates the foresight, energy, depth of knowledge, and resourcefulness that our Kessinger/Hunter team excels at delivering," says co-chairman Chuck Hunter.

Attribute part of that resourcefulness to the firm's longevity. Kessinger/Hunter first opened its doors in 1879 as a small real estate partnership. Each year, it added partners, expanded its experience, and solidified

▲ *An upscale retail space with distinctive ornamental tiles, stone, and glass, Valencia Place is the crown jewel of the Country Club Plaza. The ten-story office tower overlooks Kansas City and the Plaza and is leased and managed by Kessinger/Hunter & Company.*

its client relationships until it evolved into Greater Kansas City's leading commercial real estate company. What hasn't changed throughout the years is the firm's deep-seated commitment to each client's needs and objectives. Today Kessinger/Hunter employs more than 225 people who represent clients throughout the United States, Canada, and Europe. These clients count on the firm's knowledge, experience, and insight to achieve their objectives and receive the counsel needed to adjust to an ever-changing marketplace.

Not only is Kessinger/Hunter the region's oldest and largest commercial real estate firm, it also is one of its most diverse, equally divided into leasing and sales of industrial, commercial, and retail property, property management, and property development.

"Our diversity allows each individual who works here broad insight into how all aspects of the business operate," says Hunter. "There is constant interaction among these business groups, which allows us to predict cyclical upturns and downturns and to identify opportunities."

Since its beginnings, Kessinger/Hunter has believed in the value of encouraging associates and employees to share in the firm's success. "We have an open-door policy throughout the office," says Hunter. "Associates know they can bring their ideas to principals and management. Probably one of the great strengths of the company is the way we encourage younger associates to assume positions of responsibility and authority. We also want to encourage the next generation of leaders and producers to fulfill their ambitions, and we provide the platform from which to achieve those goals."

The real estate market may be ever-changing, but by fostering teamwork and initiative, Kessinger/Hunter consistently provides its clients with solutions to even the most complex problems. Kessinger/Hunter provides opportunities, not only for its people and for its clients, but also for dozens of communities that benefit from the company's experience, resourcefulness, and vision. ♦

> SOME COMPANIES MAY SEE THINGS ONLY AS THEY ARE, BUT KESSINGER/HUNTER & COMPANY SEES THINGS AS THEY COULD BE.

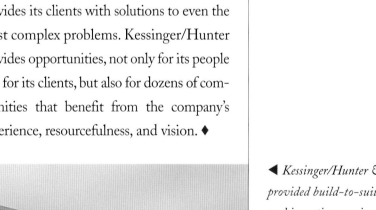

◀ *Kessinger/Hunter & Company provided build-to-suit development and incentive services for the LabOne property. It includes 270,000 square feet of corporate headquarters space and a fifty-four-acre campus.*

Khyler Wulfekotter, an avid wakeboarder, enjoys an afternoon of fun on Blue Springs Lake with parents Rayna and Tim and sister Madison, an enthusiastic novice at the sport. During the summer months, Khyler wakeboards competitively in Missouri and neighboring states; he took seventh place in a national competition in 2004, and won a Missouri state championship as well as the title of INT Junior Novice National Champion in Bakersfield, California, in 2005. INT is the International Amateur Waterski, Wakeboard, Wakeskate and Kneeboard Tour. "We love watching our kids in activities they enjoy and excel in," says Rayna. "Wakeboarding has been a wonderful opportunity for Khyler to grow his self-esteem and confidence. He is an outgoing kid and enjoys being able to hang out with new kids and families at each competition. Madison even started wakeboarding last year and is fun to watch as well. So wakeboarding is quite a family event for the Wulfekotters."

CHAPTER THREE: ENJOYING GREATER KANSAS CITY

Water, water everywhere, and all the fun you can have. Great Wolf Lodge is a first-class, year-round family destination that captures the atmosphere and adventure of the North Woods. Jace Johansen tries to tame a playful water spout before going off to get drenched under the thousand-gallon tipping water bucket that is part of the interactive Tree House Waterfort. All told, Great Wolf Lodge has a fifty-thousand-square-foot indoor entertainment area, more than enough to keep kids and adults amused.

Kansas City Southern and Kansas City: Partners in History, Partners in the Future

Railroads are synonymous with the growth of this country. Throughout history, trains have helped our country to expand to new territories, moving goods and people efficiently and safely. From its birth in 1887 as a local suburban railroad to its role today as an international carrier, Kansas City Southern (KCS) has traveled a steady track as a transportation leader with its eye on the future, and its heart in its hometown of Kansas City.

"Ever since our company's forward-thinking founder Arthur Stilwell went against what was then conventional wisdom and forged a north-south railroad, KCS has been on an unparalleled track," says Mike Haverty, chairman and CEO. "Stilwell's vision was to move agricultural products for export to a Gulf of Mexico port via the shortest possible land route. Later, he also envisioned reaching the Port of Topolabampo on Mexico's Pacific coast." Today, KCS's railroads, the Kansas City Southern Railway Company and Kansas City Southern de Mexico, operate over thirty-two hundred track miles in the United States and twenty-eight hundred track miles in Mexico, forming a rail network with primary lanes stretching from Kansas City to Mexico City and to the

▲ *A Kansas City Southern manifest train traverses a landmark bridge near Swope Park in Kansas City, Missouri.*

CHAPTER THREE: ENJOYING GREATER KANSAS CITY

PHOTO BY ERIC FRANCIS

port of Lazaro Cardenas, Veracruz, and Tampico in Mexico, and from Dallas, Texas, to Meridian, Mississippi, in the United States.

"As a single rail network, KCS will make greater investments in the coming years in cross-border transportation infrastructure, the implementation of advanced cargo tracing and tracking systems, improved border security, and world-class employee training programs," says Haverty.

KCS is also a joint venture partner in Panama Canal Railway Company (PCRC), providing trans-shipment service from the Atlantic to the Pacific oceans on a 47.6-mile railway that runs parallel to the Panama Canal. The 151-year-old line, the world's first transcontinental railroad, was recently completely modernized and serves as an efficient intermodal link for world commerce, complementing the existing transportation infrastructure provided by the Canal, the Colon free trade zone, and the Atlantic and Pacific port terminals. PCRC also operates commuter service and trains for cruise ship visitors between Colon and Panama City.

For more than a century, the visionaries, leaders, and hard-working railroaders of KCS have shared and perpetuated a unique legacy of tenacious ingenuity, helping KCS become the international railroad it is today. "As global trade between Asia and North America grows, and as more goods move between Canada, the United States, and Mexico, KCS is well positioned to handle these growing traffic volumes," adds Haverty. "Yet, while poised for dramatic growth, KCS still remains firmly rooted as Kansas City's hometown railroad." ♦

WHILE POISED FOR DRAMATIC GROWTH, KCS STILL REMAINS FIRMLY ROOTED AS KANSAS CITY'S HOMETOWN RAILROAD.

▼ *Kansas City Southern maintenance of way workers inspect the track at Grandview, Missouri.*

PHOTO BY ERIC FRANCIS

KANSAS CITY SOUTHERN | 299

Established in 1917 as the Black Musicians' Protective Union Local 627, the Mutual Musicians Foundation continues its long history of providing financial and creative support to the musicians who come through its doors. Ray Reid, shown here on the front steps, has been with the organization since 1962, when he and his band sought redress for nonpayment by a local promoter. When not singing with a variety of local music groups, including his own big band outfit, Reid helps with organizing the Mutual's Friday and Saturday after-hours jam sessions for the city's working musicians. Their jam sessions are historic; black musicians used to—and still—jam till dawn. "Being around this place is magic," says Reid. "Every musician who comes through here leaves some of themselves. If Kansas City is the Mecca of jazz, then Mutual Musicians is the Kasbah." Located at 1823 Highland Avenue in the heart of the historic 18th & Vine district, the National Historic Landmark is a true living museum, preserving and promoting the city's musical history while also serving as a venue, classroom, and rehearsal space.

CHAPTER THREE: ENJOYING GREATER KANSAS CITY

Often hailed as one of the greatest jazz musicians and possibly the best saxophonist in history, Charlie Parker was nicknamed Yardbird, later shortened to Bird, at the beginning of his career. No one knows the true origin of the nickname, but as Charlie "Bird" Parker's reputation as a superlative jazz musician grew, the name appeared in many titles of his works, and famous locations such as Birdland, the New York City nightclub, were named after him. Parker was born in 1920 in Kansas City, Kansas, and raised in Kansas City, Missouri, where this memorial to his musical genius stands today. The cast bronze image by sculptor Robert Graham is ten feet high and sits on an eight-foot base. The words "Bird Lives" etched on the pedestal pay homage to the graffiti slogan that filled the subways of New York City only hours after Parker's death at the early age of thirty-four.

PHOTO BY MARIO MORGADO

PHOTO BY DENNIS KLIM

Just as the corner of Highway 61 and Highway 49 outside Clarksdale, Mississippi, is considered the birthplace of the blues and Haight-Ashbury in San Francisco the nexus of the hippie movement and its music, so too is the corner of 18th & Vine in Kansas City an important cultural crossroads. Once the city's center for both Negro Leagues Baseball and jazz music, the area underwent a major renovation in 1997, with various museums and venues paying tribute to the city's baseball and musical heroes. Located in the area are the American Jazz Museum, the Negro Leagues Baseball Museum, the Gem Theater Cultural and Performing Arts Center, and the Horace M. Peterson III Visitors Center.

PHOTO BY MARIO MORGADO

Friends Leslie Cunningham, Beth Craig, and Paula and Robert Harr are enjoying the beautiful setting, tasty food, and good company that are a part of dinner alfresco at the Kansas City Starlight Theatre. Natives have known the venue as a landmark for live entertainment since the 1950s. However, the theatre had its start in 1925 when the Kansas City Federation of Music organized a showcase for visiting Queen Marie of Rumania. The Starlight donates 364 tickets for each Broadway show, like this production of *Grease*, staring Ryan Silverman as Danny, and Amanda Witkins as Sandy, who is surrounded by the Pink Ladies.

CHAPTER THREE: ENJOYING GREATER KANSAS CITY

PHOTO BY SCOTT INDERMAUR

It's nice to be in the middle of everything, and the Courtyard by Marriott Kansas City Country Club Plaza is just that. Even before entering their rooms, business travelers appreciate the relaxing atmosphere in the spacious lobby. The bar and market are conveniently located to welcome guests and to anticipate their needs. After a day of meetings, there is still plenty to do, because the neighborhood offers shopping, dining, and entertainment.

PHOTO BY SCOTT INDERMAUR

The Kansas City Symphony's annual Celebration at the Station attracts over thirty-five thousand people for a free concert on Memorial Day weekend. The outdoor event takes place at the historic Union Station and concludes with a spectacular fireworks display.

CHAPTER THREE: ENJOYING GREATER KANSAS CITY

It may just be a combination of hydrogen and oxygen, but when you have more than sixty acres of drenching slides and water attractions, H_2O isn't just water, it's an ocean—specifically, Oceans of Fun. Part of Worlds of Fun, this gigantic water park is located ten minutes northeast of downtown Kansas City. There are thrill rides, family attractions, and activities for little squirts. From paddling a kayak, tubing down a lazy stream, riding a floating shark, or getting drenched by a thousand-gallon bucket of water, it's summer fun for everyone. Get a group of five together and go speeding down the seventy-two-foot-high super slide or relax in the million-gallon Surf City Wave Pool. Water is said to exist in other galaxies, and that's easy to believe, because Oceans of Fun is definitely an out-of-this-world experience.

Kansas City visitors and residents alike are accustomed to seeing outdoor artworks, a tradition that started more than thirty years ago. The city recently scored a coup by hosting the inaugural exhibition of five monumental sculptures and a ten-thousand-square-foot exhibition of paintings and graphics by world-famous English artist Mackenzie Thorpe. The sculptures, each nearly fifteen feet high, were on exhibit adjacent to the J. C. Nichols Fountain in Mill Creek Park. Reportedly, Thorpe was "blown away" by the city's thriving art district. Many of Thorpe's works represent children, and the first 10 percent of sales of his art at the Leedy-Voulkos Art Center will benefit the Boys & Girls Club of Greater Kansas City.

CHAPTER THREE: ENJOYING GREATER KANSAS CITY

PHOTO BY MARIO MORGADO

Not only can guests at Hotel Phillips enjoy extra good cheer and a sparkling two-story Christmas tree during the holiday season, but also some specially priced package stays. Special dining, romance, and entertainment packages are also available throughout the rest of the year.

Kansas City's Union Station is home to the KC Rail Experience, a collection of exhibits that take visitors on an exciting journey through the nation's railroad history. In the Grandpa's Attic part of the exhibit, visitors can watch as N-gauge model trains take to the rails over miniature terrain covering one thousand square feet.

PHOTO BY MARIO MORGADO

PHOTO BY ERIC FRANCIS

PHOTO BY ERIC FRANCIS

308 | GREATER KANSAS CITY: UNLIMITED POSSIBILITIES

CHAPTER THREE: ENJOYING GREATER KANSAS CITY

Back in 1904 when horse racing was first introduced to Kansas City at the grand opening of the Elm Ridge Race Track, men with binoculars and long black cigars milled around the betting ring. Women were not allow to bet. Today, race fans at Woodlands Race Track can bet on Thoroughbred and quarter horse racing, and women have as good a chance as men to win the trifecta. Woodlands is a dual track, with a one-mile track with a straightaway for Thoroughbreds and quarter horses, and a greyhound track suitable for year-round racing. The horse racing season begins in mid-August and ends in November. Those looking for a little more action can take in the simulcast and bet on horse or dog races from all over the country. In addition to greyhounds, fans also turn out in droves for the Annual Wiener Dog Nationals, with the dachshunds flying along the track at four miles per hour. By comparison, greyhounds can run as fast as forty-five miles per hour.

During the holidays, the Kansas City Marriott Downtown stands as a glistening beacon over Barney Allis Plaza as part of the season's festivities. Some thirty thousand bulbs, which illuminate the hotel year-round, are also synchronized to flash as part of an intricate display during the kickoff celebration for the activities that take place in the downtown area.

Most serious hockey aficionados insist that real hockey is played on ice, and members of the Kansas City Stars Youth Hockey Association would probably agree. The Kansas City Amateur Hockey League Foundation formed the Stars in 1996, and currently the association has approximately four hundred players who participate on more than twenty-five teams. The program has a complete range of play levels from the four- to six-year-old Atoms, to Midget AAA for high school age. Hockey is a precise game of individual skill and team strategy, as exemplified by the eighteen-and-under Russell Stover Midget team and the Green Bay Jr. Gamblers. Russell Stover also sponsors a sixteen-and-under team. Whether cheering for a great save or a fantastic goal, players and spectators alike are passionate about the sport.

PHOTO BY ERIC FRANCIS

CHAPTER THREE: ENJOYING GREATER KANSAS CITY

Although it is now called the Last Blast of Summer, many people still think of it as the Pull the Plug event at Fairway Pool. Either way, it's a good way to end a summer of wild, wet fun. The pool is conveniently located next to Neale Paterson Park and is a prime hot-weather destination. It has two designated lap lanes, high and low diving boards, an eleven-foot slide, and a deck that's great for getting a tan or sitting in the shade playing cards. Once the season's classes and swim meets are over, everyone gets in on the tug-of-war contest. The pool's lifeguards take on the winner when Fairway Pool finally pulls the plug on summer.

PHOTO BY MARIO MORGADO

First a plate of barbecue, then a round or two of poker. That's the way they do it at Winslow's, a Kansas City staple for more than a quarter of a century. Serving up hickory-smoked meats with a choice of mouthwatering sauces, Winslow's City Market Barbecue has been winning awards since the early 1970s. Located in the historic River Market district, and open every day for lunch and dinner, Winslow's offers appetizers, sandwiches, platters, and homemade sides. Complementing the meal is Slow's Bar, where special events on weeknights include Monday and Wednesday poker games.

The word "quaff" means to drink copiously and with hearty enjoyment, and that's just part of the fun at Quaff Bar & Grill on 1010 Broadway Street in downtown Kansas City, Missouri. Apparently, the owner and his staff take the term very seriously. "We don't use our last names here; we just use Quaff as a last name," explained Joe "Quaff," owner of the establishment. This bar—family owned and operated for three generations—has been satiating the thirsts of the city's residents for sixty years. Passersby and regulars are drawn to the downtown hangout's multihued neon lights, which represent three distinct rooms: the Coors room, the Miller Lite room, and the Budweiser room.

PHOTO BY THOMAS S. ENGLAND

CHAPTER THREE: ENJOYING GREATER KANSAS CITY

For Jacob Ford (in red), playing poker at Winslow's City Market Barbecue is pretty much a break-even deal. But while he started going to Winslow's for the poker, he kept going because it's just a great neighborhood place to hang out. "You could stop going there for a few weeks, and then go back again, and the people there would remember your name," he says. A native of Kansas City, Ford moved to the River Market area and is enjoying life in a loft. "I just like the fact that Kansas City is trying really hard to bring back the downtown area," he says.

PHOTO BY THOMAS S. ENGLAND

"Our strip steaks are dry-aged for twenty-one days, so they're fabulous!" said Penny Collins, events coordinator for the Majestic Steakhouse in downtown Kansas City, Missouri. If the people who work at the Majestic enjoy the atmosphere, just think about the customers. Located in a 1911 building that was originally a bordello and bar, the venue was renovated in the mid-1980s and is listed on the National Register of Historic Places. After a sumptuous dinner, customers can listen to jazz downstairs or go upstairs to the (primarily members only) cigar club for a fine smoke in leather chairs and possibly a poker game.

PHOTO BY MARIO MORGADO

Hoefer Wysocki Builds Value in Its Communities

Ingrained in the culture of Hoefer Wysocki Architects since its establishment is the idea that architecture is as much about building relationships as it is about the buildings themselves.

"Our people and our clients are our two most valuable assets," says the firm's cofounder, principal Mitch Hoefer. "We continually seek ways to keep employees happy by providing opportunities for their growth and development, while also keeping our clients happy by realizing their vision through good planning and design."

Mitch Hoefer and principal David Wysocki formed HWA in Kansas City, Missouri, after years of project experience with national architectural firms. During that time, they accumulated a thorough knowledge of various architectural processes, practices, trends, and client issues. Over the course of their careers, the two assumed leadership roles on a variety of project types, sizes, and complexities, in over forty cities, twenty states, and two foreign countries.

When they joined forces in 1996, Hoefer and Wysocki brought to their local firm solid national and international project experience coupled with an innovative, team-oriented work environment.

▲ *The main challenge of the Liberty Hospital Medical Plaza East renovation was to unify the property's various construction and design types. The result was a seamless redesign of the entire structure that maintained efficient inpatient and outpatient movement. The firm also developed an overall master plan outlining the future growth of Liberty Hospital.*

CHAPTER THREE: ENJOYING GREATER KANSAS CITY

PHOTO ©2005 WWW.MATTNICHOLS.COM

"Mitch and Dave allow us to utilize our own past experience, project knowledge, and established relationships," says project architect Chad Ingram. "They trust their employees and the experience that we bring to the firm and, more importantly, what we can offer our clients. That trust in return opens up the opportunity for the company to serve a new market, gain new clients, and develop the company's portfolio."

As a full-service architectural, planning, and interior design firm, Hoefer Wysocki specializes in meeting the needs of clients in the commercial, health-care, judicial, government, and science and technology industries. A few notable projects include the Federal Government Campus in Birmingham, Alabama; the Saunders County Health System in Wahoo, Nebraska; and the Shawnee Justice Center & Fire Station No. 2 in Shawnee, Kansas. Hoefer Wysocki has also developed numerous commercial and retail office projects throughout California, Texas, and Florida.

Another of the firm's founding beliefs is that the success of every project comprises three key elements: client service, innovative design, and sound project management. These three elements are best exemplified in the firm's use of the charrette design process: an on-site, interactive, participatory design workshop that involves the client, stakeholders, design team, engineers, and in some cases members of the community. The goal of the charrette is to encourage productive brainstorming that results in a thorough understanding of the client's vision and project requirements. Not only is the charrette a creative and efficient way to create a preliminary design, it builds trust between team members and invests each of them with a sense of ownership in the project.

That kind of innovative thinking and dedication to client service first attracted

(continued on page 316)

ARCHITECTURE IS AS MUCH ABOUT BUILDING RELATIONSHIPS AS IT IS ABOUT THE BUILDINGS THEMSELVES.

▼ *Contrasting bands of brick and cast stone give the Pinnacle II and III buildings at Tomahawk Creek a strong presence in the corporate campus. The "L" shape created by the two buildings allows for multiple corner offices with views to the surrounding woods.*

PHOTO BY ALAN S. WEINER

HOEFER WYSOCKI ARCHITECTS, LLC

PHOTO BY ALAN S. WEINER

(continued from page 315)

principal and director of design Kevin Berman to the firm. "After meeting David and Mitch and learning about their depth and breadth of national project experience, I realized that I could fulfill my goal of designing state-of-the-art buildings and world-class architecture right here in Kansas City," he says. As the firm's first employee, Berman has watched the firm grow from two to forty employees and establish not only close relationships with employees and clients, but also with numerous developers, construction firms, and subconsulting firms. "This growth is a

▲ When premier Kansas City jeweler Tivol Jewelers decided to double the size of its existing store and create a new location at Briarcliff development, it called on Hoefer Wysocki to oversee both projects. The highlight of the new location is its museum-inspired interior, which displays Tivol's jewelry like works of art in cases created from pear and maple wood with black granite accents.

◀ Like the building's exterior, the lobby of the Pinnacle Corporate Centre at Tomahawk Creek in Leawood showcases the firm's sophisticated use of contrasting materials, creating an elegant introduction to the seventy-five-thousand-square-foot office building.

CHAPTER THREE: ENJOYING GREATER KANSAS CITY

◀ *Because they understand the "science" behind the design of medical office buildings, Hoefer Wysocki designed the interior of Corporate Medical Plaza to complement the exterior. The warm color palette provides a distinct character that breaks away from the mold of a traditional medical office building setting. The high-quality, beautiful finishes, such as rich woods and granites in the lobbies and common areas, create a comforting and pleasing environment that begins as you walk in the main doors of the building.*

direct result of consistently performing on every project and exceeding client expectations," he says.

Over the years, HWA's dedication to its employees and clients has earned it several notable business and design awards. Consistently ranked as one of Kansas City's fastest-growing architectural firms, it was also named Small Business of the Year 2006 by the Greater Kansas City Chamber of Commerce and one of The Chamber's Top 10 Small Businesses for 2005. Award-winning designs include the FBI Birmingham Field Office, which was named Best Public Building Project of the Year 2005 by South Central construction, and Riverside, Missouri's city hall, which received the American Public Works Association's Project of the Year award in 2002.

Building solid working relationships with employees and clients is one way Hoefer Wysocki is helping communities prosper. Lending a helping hand to those less fortunate is another. Whether donating five thousand cans of food to the Harvester food bank, participating in the yearly Kansas City Public Television fund drive, or painting the streets of Kansas City as part of the annual La Strada dell Arte festival, firm employees are actively making a difference. Each year HWA also closes its offices entirely to work on another kind of building: a Habitat for Humanity home for a local family. By putting its own projects and clients on hold for just one day, HWA makes for someone in need a difference that lasts a lifetime. ♦

▼ *Hoefer Wysocki's masterplan of the Olathe Public Safety Facility campus integrated the existing aesthetic of downtown Olathe with new state-of-the-art facilities to serve the municipal courts, police department, and fire administration.*

HOEFER WYSOCKI ARCHITECTS, LLC | 317

CHAPTER THREE: ENJOYING GREATER KANSAS CITY

It's a world of color and fascination at the Hallmark Visitors Center. Housed in Crown Center, the jewel of Kansas City's downtown, the center pays tribute to the story of J. C. Hall, founder of the world's greeting-card giant. In addition to displays that tell the history of Hallmark, the center's interactive exhibits give artists, and the artistically challenged, a chance to explore and experience the same kind of creative thinking that guides the company today.

Greater Kansas City Featured Companies

AMC Entertainment Inc.
920 Main Street
Kansas City, Missouri 64105
816.221.4000
www.amctheatres.com

Attraction–Theatres (pp. 220–221)
One of the country's oldest, largest, and most innovative exhibition companies, AMC Entertainment Inc. was established in Kansas City in 1920. Today it provides more than 240 million guests each year with the best possible out-of-home entertainment experience.

American Century Investments
4500 Main Street
Kansas City, Missouri 64111
800.345.2021
www.americancentury.com

Investment Company–Stock Broker (pp. 132–135)
Established by James E. Stowers in 1958, this privately held investment management firm helps millions of individuals, corporations, and institutions meet their financial goals. Defined by innovation and high ethical standards, American Century Investments continually receives accolades from industry rating companies like Morningstar and *Fortune* magazine.

American Digital Security
1205 West College Street
Liberty, Missouri 64068
816.415.4237 or 888.833.4237
www.adscc.tv

Security Company (pp. 50–51)
American Digital Security specializes in the design, system integration, installation, and service of closed circuit television (CCTV), digital/network video surveillance, card access control, IP products solutions, door intercom systems, and intrusion detection systems for commercial clients, with holdings ranging from a single location to multiple-property chain stores.

Applebee's International, Inc.
4551 West 107th Street
Overland Park, Kansas 66207
913.967.8117
www.applebees.com

Restaurant–Corporate Office (pp. 254–255)
Applebee's Neighborhood Grill & Bar is the largest casual dining concept in America. With headquarters in Overland Park, Kansas, it has restaurants in forty-nine states and seventeen countries. Applebee's restaurants reflect their neighborhoods, from décor, to specialty menu items, to the staff, and their support of local organizations.

Aquila
20 West Ninth Street
Kansas City, Missouri 64105
816.467.3520
www.aquila.com

Utility (pp. 64–66)
Aquila Inc. is an electricity and natural gas utility based in Kansas City, Missouri, that has distribution facilities and serves 1 million customers in Colorado, Iowa, Kansas, Missouri, and Nebraska. The company is well-known for its knowledgeable, helpful staff and its philanthropic activities.

Armstrong Teasdale LLP
2345 Grand Boulevard, Suite 2000
Kansas City, Missouri 64108
816.221.3420
www.armstrongteasdale.com

Law Firm (pp. 186–187)
Armstrong Teasdale is a leading law firm that delivers sophisticated legal advice and exceptional client service. With over 265 attorneys practicing in eleven national and international locations, Armstrong Teasdale serves a dynamic national and international client base in virtually every area of law. Whether an issue is local or global, practice-area-specific or industry-related, Armstrong Teasdale provides each client with an invaluable combination of legal and practical advice capable of securing optimal results.

Bank Midwest
1100 Main Street
Suite 416
Kansas City, Missouri 64196
816.471.9800
www.bankmw.com

Financial Institution (pp. 154–156)
Since its founding in 1970, Bank Midwest has become one of the largest and most successful privately held financial institutions in the region. This success has been due in large part to the bank's owners and executives continually looking to the future and being prepared with banking products and services to meet evolving customer needs.

Bayer CropScience
8400 Hawthorn Road
Kansas City, Missouri 64120-0013
816.242.2000
www.bayercropscience.com

Manufacturer (p. 46)
The state-of-the-art Bayer CropScience manufacturing facility in Kansas City, Missouri, provides innovative crop protection technologies that help farmers fight damaging pests, plant diseases, and weeds, and thereby achieve abundant and wholesome harvests in an affordable, sustainable, and environmentally sound way.

Bayer HealthCare LLC
12707 Shawnee Mission
Shawnee, Kansas 66216
913.268.2000
www.animalhealth.bayerhealthcare.com

Manufacturer–Pharmaceuticals (p. 47)
Bayer HealthCare's Animal Health Division provides best-in-class therapies for both companion animals and livestock. With cornerstone products like Advantage®, K9 Advantix®, Baytril®, and Legend®, the company continues to seek new treatments to improve the health and well-being of animals.

BKD, LLP
120 West Twelfth Street, Suite 1200
Kansas City, Missouri 64105
816.221.6300
www.bkd.com

CPA Firm (pp. 208–210)
BKD, LLP, is one of the ten-largest CPA and advisory firms in the country and the second-largest in Kansas City. We help clients go beyond their numbers by applying our technical expertise, unmatched client service, and disciplined delivery of solutions to their management and financial needs.

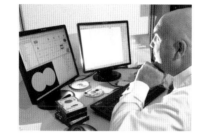

Black & Veatch Corporation
8400 Ward Parkway
Kansas City, Missouri 64114
913.458.2000
www.bv.com

Engineering Firm–Diverse (pp. 158–161)
Whether building the world's fourth-largest hydropower station in China, expanding the globe's largest nitrogen plant in Mexico, or restoring telecommunications in storm-ravaged New Orleans, Black & Veatch Corporation delivers on its promise: Building a World of Difference®. Founded in 1915 in Kansas City, Missouri, Black & Veatch is a leading global engineering, consulting, and construction company, specializing in infrastructure development in energy, water, telecommunications, management consulting, federal, and environmental markets.

Blackwell Sanders Peper Martin LLP
4801 Main Street
Suite 1000
Kansas City, Missouri 64112
816.983.8000
www.blackwellsanders.com

Law Firm (pp. 146–148)
Since 1916, this commercial law firm has provided legal solutions to meet the complex business and regulatory challenges of its clients. Today, it has grown into one of the Midwest's leading commercial law firms, with clients that span the nation and the globe.

Blue Cross and Blue Shield of Kansas City
2301 Main Street
Kansas City, Missouri 64108
816.395.2222
www.bcbskc.com

Insurance–Health (pp. 54–57)
Blue Cross and Blue Shield of Kansas City is the Kansas City area's most prominent health insurer, offering affordable, accessible options for large and small employer groups, individuals, and families.

Burns & McDonnell
9400 Ward Parkway
Kansas City, Missouri 64114
816.333.9400
www.burnsmcd.com

Engineering Firm, Construction, Environmental, and Consulting (pp. 122–125)
Burns & McDonnell is a Kansas City–based engineering, architecture, construction, environmental, and consulting services firm that has completed projects in every state and thirty-three countries. Founded in 1898, the firm has twenty-three hundred employees in offices nationwide.

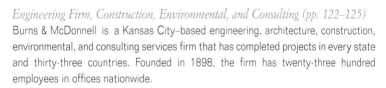

Chase Suites Hotels
9900 NW Prairie View Road
Kansas City, Missouri 64153
816.891.9009
www.woodfinsuitehotels.com

Hotel (pp. 258–259)
A tradition in all-suite hospitality services for business and pleasure travelers, Chase Suites Hotels believes in making its guests feel right at home. Stylish and comfortable accommodations, convenient locations, and a host of complimentary services combine to create a top hotel of choice in Kansas City and across the nation.

Children's Mercy Hospitals
2401 Gillham Road
Kansas City, Missouri 64108
816.234.3000
www.childrensmercy.org

Hospital–Medical Center (pp. 60–61)
Children's Mercy Hospitals and Clinics began in 1897 as a one-room haven for a five-year-old crippled child. Today, named by *Child* magazine as one of the top children's hospitals in America, Children's Mercy is a state-of-the-art, 314-bed system that also operates outpatient clinics throughout the region. The Children's Mercy staff treats the full range of childhood illnesses, from outpatient surgery to chronic disease and critical care of children from birth to adolescence.

City of Blue Springs
1600 NE Coronado Drive
Blue Springs, Missouri 64014
816.228.0208
www.bluespringsedc.com

Government–City (pp. 190–192)
Blue Springs, Missouri, is a thriving Kansas City suburb undergoing an exciting stage of development and growth that will complement its existing small community character, attractive quality of life, available and productive workforce, and incentive support for existing and new business investment and job creation.

Colliers Turley Martin Tucker
2600 Grand, Suite 1000
Kansas City, Missouri 64108
816.221.2200
www.ctmt.com

Real Estate–Commercial (p. 40)
Colliers Turley Martin Tucker is the dominant commercial real estate firm in the Central United States, with more than twelve hundred associates serving client needs in eight regional offices in Missouri, Ohio, Indiana, Minnesota, and Tennessee.

Courtyard by Marriott Country Club Plaza
4600 J. C. Nichols Parkway
Kansas City, Missouri 64112
816.285.9760
www.marriott.com

Hotel (pp. 224–226)
The historic Courtyard by Marriott is located on the beautiful Country Club Plaza. It incorporates the latest technology in its rooms, and has 924 square feet of meeting space, plus ballrooms ranging from 580 to 1,400 square feet. The hotel is in the heart of a dining, entertainment, and shopping district as well as being convenient to all major businesses and only minutes from the Kansas City International Airport.

Ewing Marion Kauffman Foundation
4801 Rockhill Road
Kansas City, Missouri 64110
816.932.1000
www.kauffman.org

Nonprofit–Foundation (pp. 26–28)
The Ewing Marion Kauffman Foundation seeks to create a society of independent, productive citizens through grants and initiatives focused on advancing entrepreneurship and improving education.

Executive AirShare
150 Richards Road, Suite 103
Kansas City, Missouri 64116
816.221.7200
www.execairshare.com

Transportation (pp. 230–231)
Executive AirShare is a regional, fractional aircraft company that sells ownership shares in private aircraft. "Once you own a share of one of our aircraft, you tell us when and where you want to go, and we take care of the rest. It's as close as you can get to owning your own plane without the huge investment or the hassle," says Bob Taylor, president and CEO.

Fogel-Anderson Construction Co.
1212 East Eighth Street
Kansas City, Missouri 64106
816.842.6914
www.fogel-anderson.com

Contractor–General (pp. 164–165)
Specialists in general contracting, design/build, and construction management, Fogel-Anderson Construction Co. has built an excellent reputation in the commercial and industrial industry since 1917. Family-owned for four generations, the organization's management continues the traditions of valuing employees and providing excellent service for customers.

Garmin International, Inc.
1200 East 151st Street
Olathe, Kansas 66062
913.397.8200
www.garmin.com

Manufacturer–GPS Devices (pp. 288–291)
Garmin designs, manufactures, and markets GPS navigation and communications equipment for general aviation, marine, automotive, outdoor, mobile electronics, and personal fitness. With facilities in the United States, Europe, and Asia, the Garmin brand is known for innovative, award-winning products that set the standard in the industry.

Greater Kansas City Chamber of Commerce
911 Main Street, Suite 2600
Kansas City, Missouri 64105
816.221.2424
www.kcchamber.com

Gould Evans
4041 Mill Street
Kansas City, Missouri 64111
816.931.6655
www.gouldevans.com

Architectural Firm (pp. 170–173)
Gould Evans designs places for people to live, learn, work, and play. The firm's 220 talented, committed associates provide architecture, urban planning and design, landscape architecture, interior design, and graphic design services from seven offices across the country, including three in the Greater Kansas City region.

Chamber of Commerce (pp. 128–129)
The Greater Kansas City Chamber of Commerce serves as a facilitator of regional cooperation. With members in both Missouri and Kansas, the organization takes the lead in public policy issues, working with business leaders and legislators toward positive change. The Chamber's programs and services provide members with avenues for networking, development, and greater exposure.

Greater Kansas City Community Foundation
1055 Broadway, Suite 130
Kansas City, Missouri 64105
816.842.0944
www.gkccf.org

Nonprofit–Foundation (p. 74)
The seventh-largest community foundation in the country, the Greater Kansas City Community Foundation for more than twenty years has been connecting donors to the needs in the community they care about. The Community Foundation, including its five affiliates within the region, is recognized as a national leader in making sure every philanthropic investment returns the greatest emotional, civic, and financial benefit possible.

Hallmark Cards, Inc.
2501 McGee Trafficway
Kansas City, Missouri 64108
1-800-HALLMARK
www.hallmark.com

Manufacturer–Greeting Cards (pp. 244–245)
Founded in 1910 by Joyce C. Hall, privately held Hallmark Cards Inc. is the industry leader in personal expression. Each year, Hallmark and its subsidiaries produce more than forty thousand products in more than thirty languages distributed in one hundred countries. Of the seventeen thousand full-time employees worldwide, approximately four thousand work at the Kansas City headquarters.

HDR Engineering, Inc.
4425 Main Street, Suite 1000
Kansas City, Missouri 64111
816.360.2724
www.hdrinc.com

Engineering Firm–Diverse (p. 140)
HDR is a national, employee-owned architectural, engineering, and consulting firm that excels at managing complex projects and solving challenges for its clients. HDR specializes in building, shaping, and serving its local communities.

The HNTB Companies
715 Kirk Drive
Kansas City, Missouri 64105
816.472.1201
www.hntb.com

Architecture/Engineering Firm (pp. 278–279)
The HNTB Companies is an employee-owned organization of infrastructure firms known and respected for its work in transportation, bridges, aviation, architecture, and urban design and planning. The company's subsidiaries, which employ more than three thousand people nationwide, consist of HNTB Corporation, an engineering, planning, and construction management firm; HNTB Architecture Inc., providing specialized buildings services; and HNTB Federal Services, serving federal-sector clients.

Hoefer Wysocki Architects, LLC
612 West Forty-seventh Street
Suite 300
Kansas City, Missouri 64112
816.221.0606
www.hwa.net

Architectural Firm (pp. 314–317)
This full-service, client-oriented architectural, planning, and interior design firm takes a team-oriented approach to developing innovative, cost-effective projects for the commercial, health-care, judicial, government, and science and technology industries.

Hotel Phillips
106 West Twelfth Street
Kansas City, Missouri 64105
816.221.7000
www.hotelphillips.com

Hotel (pp. 270–273)
Built in 1931, this beautifully renovated art deco masterpiece epitomizes the spirit of Kansas City, seamlessly mixing old-world elegance, attentive service, and modern-day accommodations and amenities.

JE Dunn Construction
929 Holmes
Kansas City, Missouri 64106
816.474.8600
www.jedunn.com

Contractor–General (pp. 180–183)
JE Dunn Construction is a family-owned business founded in 1924. With headquarters in Kansas City, Missouri, the company specializes in construction management, program management, general construction, and design/build projects of every size. The sixth-largest general building contractor in the United States, the JE Dunn Construction Group comprises six construction companies with offices in seventeen locations across the country.

Johnson County Community College
12345 College Boulevard
Overland Park, Kansas 66210-1299
913.469.8500
www.jccc.edu

School–Community College (p. 96)
Johnson County Community College (JCCC) is Kansas's third-largest institution of higher education and the largest of the state's community colleges. Founded in 1969, JCCC provides developmental, transfer, and career education and enjoys transfer agreements with one hundred nearby colleges and universities.

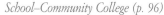

Johnson County Government
Johnson County Square
111 South Cherry
Olathe, Kansas 66061
913.715.0725
www.jocogov.org

Government–County (pp. 70–71)
One of the first counties in Kansas, Johnson County is an up-and-coming suburban community comprising twenty cities. The Johnson County Government, with over thirty-nine hundred employees, continually works with residents to responsibly develop a region that attracts businesses and families to the good life.

Kansas City Area Development Council
2600 Commerce Tower
Kansas City, Missouri 64105
816.374.5620
www.thinkkc.com

Non-Profit (pp. 30–31)
This private, nonprofit economic development organization not only works to attract new business to the eighteen-county, bi-state Kansas City region, but also promotes the area as a unified region through its OneKC and ThinkKC campaigns.

Kansas City Marriott Downtown
200 West Twelfth Street
Kansas City, Missouri 64105
816.421.6800
www.kcmarriott.com

Hotel (pp. 248–251)
The Kansas City Marriott Downtown comprises two connected structures that contrast architecturally, but that together welcome guests with a long-standing heritage of hospitality. The hotel is Kansas City's premier downtown site for accommodations and events.

Kansas City Power & Light
1201 Walnut Street
Kansas City, Missouri 64141-9679
816.556.2372
www.kcpl.com

Utility (pp. 42–44)
Kansas City Power & Light, founded in 1882, provides energy for more than half a million households in twenty-four counties of western Missouri and eastern Kansas.

Kansas City Regional Association of REALTORS®
11150 Overbrook Road, Suite 100
Leawood, Kansas 66211
913.498.1100
www.kcrealtorlink.com

Association (pp. 92–93)
The Kansas City Regional Association of REALTORS® is a not-for-profit trade association representing approximately ten thousand members in Greater Metropolitan Kansas City and surrounding counties in Kansas. Its mission is to advance the professional and business interests of REALTORS® through advocacy, services, and communication.

Kansas City Southern
427 West Twelfth Street
Kansas City, Missouri 64105
816.983.1303
www.kcsouthern.com

Transportation (pp. 298–299)
Kansas City Southern (KCS) is a transportation holding company that has railroad investments in the United States, Mexico, and Panama. Its primary U.S. holding is the Kansas City Southern Railway Company. Its international holdings include Kansas City Southern de Mexico, and a 50 percent interest in Panama Canal Railway Company. KCS's North American rail holdings and strategic alliances are primary components of a NAFTA Railway system, linking the commercial and industrial centers of the United States, Mexico, and Canada.

324 | GREATER KANSAS CITY: UNLIMITED POSSIBILITIES

Kansas City Symphony
1020 Central #300
Kansas City, Missouri 64105
816.471.1100
www.kcsymphony.org

Nonprofit–Symphony (pp. 264–266)
During its forty-two-week season, the Kansas City Symphony (Michael Stern, music director) performs classical, pops, educational, and family series concerts and takes music into the community through educational programs, ensemble presentations, and free public concerts. The symphony also performs for the Lyric Opera of Kansas City and the Kansas City Ballet.

Kansas City University of Medicine and Biosciences
1750 Independence Avenue
Kansas City, Missouri 64106
800.234.4847
www.kcumb.edu

School–University/Medical (pp. 34–37)
Kansas City University of Medicine and Biosciences provides high-quality educational programs for physicians, scientists, and researchers. Its facilities and charitable activities offer significant opportunities for hands-on research, patient care, and community involvement.

KCMO Water Services Department
4800 East Sixty-third Street
Kansas City, Missouri 64130
816.513.0123
www.kcmo.org/water

Utility (pp. 76–77)
Serving Kansas City since 1895, KCMO Water Services Department is committed to providing safe, quality drinking water for city residents. The department is also responsible for treating wastewater in an environmentally friendly manner, protecting the city from flooding, and ensuring stormwater quality.

Kessinger/Hunter & Company
2600 Grand Boulevard, Suite 700
Kansas City, Missouri 64108
816.842.2690
www.kessingerhunter.com

Real Estate–Commercial (pp. 294–295)
Established in 1879, Kessinger/Hunter is the oldest commercial real estate company in Greater Kansas City. That depth and breadth of experience enable the firm to meet the most complex of its clients' real estate needs, as well as seek out new investment opportunities.

Key Companies & Associates LLC
1331 Swift
Kansas City, Missouri 64116
816.471.1333
info@jjkeyco.com

Entrepreneur (pp. 202–203)
Key Companies & Associates LLC (founders of S&M NuTec, Makers of Greenies®) is a venture of Joe and Judy Roetheli, entrepreneurs who credit their success to the supportive business environment found in Kansas City.

Lathrop & Gage L.C.
2345 Grand Boulevard
Suite 2800
Kansas City, Missouri 64108-2612
816.292.2000
www.lathropgage.com

Law Firm (pp. 142–143)
A leading midwestern full-service law firm, Lathrop & Gage L.C. ranks No. 152 on the *National Law Journal's* list of the country's largest law firms, with 260 attorneys in ten offices nationwide. In 2006, Chambers USA ranked Lathrop & Gage's corporate, litigation, real estate, and labor and employment practices among the best in the Midwest, noting that the firm is populated with "excellent legal advisors with a strong commitment to customer service."

Lewis, Rice & Fingersh, L.C.
One Petticoat Lane
1010 Walnut, Suite 500
Kansas City, Missouri 64106
816.421.2500
www.lewisrice.com

Law Firm (pp. 176–177)
Lewis, Rice & Fingersh, L.C., is a law firm that combines personalized representation with big-firm capabilities to deliver solutions for clients across the nation. While the firm's practice encompasses a full scope of legal expertise, its Kansas City office specializes largely in matters involving commercial real estate, corporate law, and litigation.

Mark One Electric Company Inc.
909 Troost
Kansas City, Missouri 64106
816.842.7023

Contractor–Electrical (pp. 84–85)
Since 1974, this family-owned and -operated electrical contractor has provided quality electrical design and construction services for some of the Midwest's largest industrial and commercial construction projects.

Midwest Research Institute
425 Volker Boulevard
Kansas City, Missouri 64110
816.360.1943
www.mriresearch.org

Metropolitan Community College
3200 Broadway
Kansas City, Missouri 64111
816.759.1000
www.mcckc.edu

School–Community College (pp. 88–89)
Since 1915, Metropolitan Community College has served the educational needs of a diverse population, providing students with the education to enhance their future and the future of Greater Kansas City.

Research (p. 110)
Founded in 1944, Midwest Research Institute's scientific research encompasses advanced work in national defense, health sciences, agriculture and food safety, engineering, and energy. MRI, headquartered in Kansas City, operates offices and laboratories in Florida and Maryland, and also manages the U.S. Department of Energy's National Renewable Energy Laboratory in Colorado.

Park University
8700 NW River Park Drive
Parkville, Missouri 64152
816.584.6211
www.park.edu

School–University (pp. 100–101)
Park University has forty-three campuses nationwide and online, plus four campuses in the Kansas City area. More than 60 percent of the students are active-duty military personnel, military dependents, retired military, or associated with the Department of Defense. The school is on the cutting edge of technology, and in 2006 was ranked second by U.S. News and World Report for providing online degrees.

Polsinelli Shalton Flanigan Suelthaus PC
700 West Forty-seventh Street
Suite 1000
Kansas City, Missouri 64112
816.753.1000
www.polsinelli.com

Law Firm (pp. 150–151)
Based in Kansas City and with offices in St. Louis, Chicago, New York, Washington, D.C., Overland Park and Topeka, Kansas, and Edwardsville, Illinois, this energetic first-generation law firm is dedicated to providing its clients with innovative solutions for a variety of business, real estate, financial, and litigation needs.

Shook, Hardy & Bacon LLP
2555 Grand Boulevard
Kansas City, Missouri 64108
800.821.7962
www.shb.com

Law Firm (pp. 198–199)
Shook, Hardy & Bacon is a full-service, international firm founded in 1889. With its largest office in Kansas City, and nine other offices strategically located throughout the world, SHB serves a diversified client base with a wide range of practice groups.

Sprint
6200 Sprint Parkway
Overland Park, Kansas 66251
913.794.3415
www.sprint.com

Telecommunications (pp. 236–239)
Sprint has stood at the forefront of the telecommunications industry since its establishment in Abilene, Kansas, in 1899. Today, it continues to provide some of the most innovative, life-enhancing communications products and services for both business and personal use.

SwopeCommunity Enterprises
4001 Blue Parkway, Suite 270
Kansas City, Missouri 64130
www.swopecommunity.org

Health-Care Services (pp. 104–107)
Swope Community Enterprises represents a complementary range of essential community services to improve the physical, behavioral, and economic health of the underserved in targeted urban areas. It provides primary health care, behavioral health care, commercial and economic development planning and housing, and commercial development.

Top Innovations, Inc.
6655 Troost Avenue
Kansas City, Missouri 64131
816.584.9700
www.topinnovations.com

Manufacturing Representative (pp. 194–195)
Top Innovations Inc. specializes in developing consumer products that make life easier. Since 1987, the company, based in Kansas City, Missouri, has operated its international business with a mission and a commitment to provide customers with products that bring solutions to their everyday needs.

Truman Medical Centers
2301 Holmes Street
Kansas City, Missouri 64108
816.404.1000
www.trumed.org

Hospital–Medical Center (p. 82)
Truman Medical Centers (TMC), a nonprofit, two-hospital system, was named by Solucient as a Top 100 Hospital due to its high ratings for improved patient outcomes and financial performance. Together, the two hospitals (TMC Hospital Hill and TMC Lakewood) and the center of Behavioral Health provide a full-service network of health care for the growing Kansas City urban and suburban communities.

The University of Kansas Hospital
3901 Rainbow Boulevard
Kansas City, Kansas 66160
903.588.1441
www.kumc.edu

Hospital–Medical Center (pp. 22–23)
The University of Kansas Hospital, in its second century of service, has made a dramatic turnaround to once again become the region's premier academic medical center, providing a full range of both inpatient and outpatient services. Through its affiliation with the University of Kansas Schools of Medicine, Nursing, and Allied Health, the hospital also participates in some of the nation's most exciting research and clinical trials.

WDS Marketing & Public Relations
6524 West Forty-ninth Street
Mission, Kansas 66202
913.362.4541
www.wdspr.com

Public Relations Agency (p. 282)
WDS Marketing & Public Relations was founded in 1991 by Becky S. Wilson. This award-winning agency provides concepts and follow-through for marketing strategies, public relations campaigns, media relations programs, innovative brand building, and creative business events.

Many thanks for your support!

About the Publisher

Greater Kansas City: Unlimited Possibilities was published by Bookhouse Group, Inc., under its imprint of Riverbend Books. What many people don't realize is that in addition to picture books on American communities, we also develop and publish institutional histories, commemorative books of all types, contemporary books, and others for clients across the country.

Bookhouse has developed various types of books for prep schools from Utah to Florida, colleges and universities, country clubs, a phone company in Vermont, a church in Atlanta, hospitals, banks, and many other entities. We've also published a catalog for an art collection for a gallery in Texas, a picture book for a worldwide Christian ministry, and a book on a priceless collection of art and antiques for the Atlanta History Center.

These beautiful and treasured tabletop books are developed by our staff as turnkey projects, thus making life easier for the client. If your company has an interest in our publishing services, do not hesitate to contact us.

Founded in 1989, Bookhouse Group is headquartered in a renovated 1920s tobacco warehouse in downtown Atlanta. If you're ever in town, we'd be delighted if you looked us up. Thank you for making possible the publication of *Greater Kansas City: Unlimited Possibilities.* ❖

BOOKHOUSE
GROUP, INC.

Banks ❖ Prep Schools ❖ Hospitals ❖ Insurance Companies ❖ Art Galleries ❖ Museums ❖ Utilities
❖ Country Clubs ❖ Colleges ❖ Churches ❖ Military Academies ❖ Associations

Kansas City Editorial Team

Kimberly Fox DeMeza, Writer, Roswell, Georgia. Combining business insight with creative flair, DeMeza writes to engage the audience as well as communicate the nuances of the subject matter. While officially beginning her career in public relations in 1980 with a degree in journalism, and following in 1990 with a master's in health management, writing has always been central to her professional experience. From speech-writing to corporate brochures to business magazine feature writing, DeMeza enjoys the process of crafting the message. Delving into the topic is simply one of the benefits, as she believes every writing opportunity is an opportunity to continue to learn.

Rena Distasio, Writer, Tijeras, New Mexico. Freelance writer Rena Distasio contributes articles and reviews on a variety of subjects to regional and national publications. She also edits two magazines focused on living and playing in the Four Corners region. In her spare time she and her husband and two dogs enjoy the great outdoors from their home in the mountains east of Albuquerque.

Grace Hawthorne, Writer, Atlanta, Georgia. Starting as a reporter, she has written everything from advertising for septic tanks to the libretto for an opera. While in New York, she worked for Time-Life Books and wrote for *Sesame Street*. As a performer, she has appeared at the Carter Presidential Center, Callanwolde Fine Arts Center, and at various corporate functions. Her latest project is a two-woman show called *Pushy Broads and Proper Southern Ladies*.

Regina Roths, Writer, Andover, Kansas. Roths has written extensively about business since launching her journalism career in the early 1990s. Her prose can be found in corporate coffee-table books nationwide as well as on regionally produced Web sites, and in print and online magazines, newspapers, and publications. Her love of industry, history, and research gives her a keen insight into writing and communicating a message.

Gail Snyder, Writer, Woodstock, Georgia. Snyder is a writer and editor with twenty years of experience in corporate communications and publishing. She has edited or written articles focusing on corporate management strategies, published articles in a number of trade magazines and journals, and edited both fiction and nonfiction books. Gail enjoys explaining material to an audience in a way that reveals how any subject can be interesting. She earned her bachelor's degree in journalism from Georgia State University, where she went on to complete her master's in communications. Currently she works as a freelance and contract writer and editor.

Thomas S. England, Photographer, Decatur, Georgia. England grew up internationally, graduated from Northwestern University, and began photography as a newspaper photographer in the Chicago area. He began freelancing for *People* magazine in 1974. Since then he has taken assignments from national magazines and corporations, specializing in photographing people on location. He lives in Decatur, Georgia, with Nancy Foster, a home renovator, and their dog Chessey. More of his photographs may be viewed online at www.englandphoto.com.

Eric Francis, Photographer, Omaha, Nebraska. Francis was born and raised in Nebraska. Early on, he honed his skills freelancing for local newspapers, magazines, and commercial clients. Francis now also works regularly for some of the nation's largest and best-known magazines, newspapers, and wire services, covering news, features, and sports. He continues to make his home in Omaha with his fiancé, Shannon, and their four children.

Dennis Keim, Photographer, Huntsville, Alabama. Keim is a photographer with over twenty-five years of experience producing creative imagery for the editorial, advertising, and corporate communities. His editorial work has been featured in regional, national, and international publications. Notable in his professional career was his employment as an Aerospace Photojournalist by NASA (National Aeronautics and Space Administration) from 1976 to 2000 at the Marshall Space Flight Center. Dennis is a member of ASMP and currently maintains a commercial studio specializing in corporate, editorial, and stock photography. In addition he is an exhibited and published fine arts photographer. You can view more of his work at www.dk-studio.com.

Scott Indermaur, Photographer, Lawrence, Kansas. Indermaur's editorial and corporate assignments have taken him from the smallest rural communities to the world's most urban environments. His gift lies in discovering the familiar in the exotic and the remarkable in the ordinary. Whether he's capturing a fleeting moment in history or cutting to the essence of a portrait, Scott tells the story in a language everyone understands. When not photographing, Scott, who along with his wife and son are all Tae Kwon Do black belts, also enjoys wonderful food and wine, meeting new people, traveling, music, and sailing. You can view his images and contact him at www.siphotography.com.

Mario Morgado, Photographer, Greenwich, Connecticut. Morgado was born in Cuba and somehow managed to have a happy childhood while growing up in Elizabeth, New Jersey. He spent his formative years at the Guggenheim Museum and the right-field bleachers of Yankee Stadium. His work has appeared in numerous publications including *The New York Times*, *Vermont Life*, *New York* magazine, and *Boston* magazine. Mario's work can be seen at www.mariomorgado.com.

Alan S. Weiner, Photographer, Portland, Oregon. Weiner travels extensively both in the United States and abroad. Over the last twenty-three years his work has appeared regularly in *The New York Times*. In addition, his pictures have been published in *USA Today* and in *Time*, *Newsweek*, *Life*, and *People* magazines. He has shot corporate work for IBM, Pepsi, UPS, and other companies large and small. He is also the cofounder of the Wedding Bureau (www.weddingbureau.com). Alan has worked in every region of the country for Riverbend Books. His strengths are in photojournalism.